Saudi America
The Truth About Fracking and How It's Changing the World

COLUMBIA GLOBAL REPORTS
NEW YORK

Saudi America
The Truth About Fracking and How It's Changing the World

Bethany McLean

Canada

Marcellus
Formation

Atlantic Ocean

© 2018 Jeffrey L. Ward

Saudi America
The Truth About Fracking and How It's Changing the World
Copyright © 2018 by Bethany McLean
All rights reserved

Published by Columbia Global Reports
91 Claremont Avenue, Suite 515
New York, NY 10027
globalreports.columbia.edu
facebook.com/columbiaglobalreports
@columbiaGR

Library of Congress Control Number: 2018949784
ISBN: 9780999745441

Book design by Strick&Williams
Map design by Jeffrey L. Ward
Author photograph by Miranda Sita

Printed in the United States of America

Saudi America
The Truth About Fracking
and How It's Changing
the World

CONTENTS

Introduction

In the late afternoon on New Year's Eve, 2015, the *Theo T*, a hulking, dark gray Bahamian tanker, gingerly maneuvered in the light rain through a channel from the North Beach Terminal at the port of Corpus Christi in southern Texas into the open waters of the Atlantic Ocean. The *Theo T* was fully loaded with American crude oil that had been drilled by Conoco Phillips in Texas's Eagle Ford shale, a rock formation deposited over sixty-five million years ago that became, in the modern age of fracking, one of the most prolific oil fields in the United States. The oil had traveled one hundred miles through pipeline owned by San Antonio-based NuStar Energy. Twenty days after the *Theo T* disembarked from Corpus Christi, the oil would arrive at Marseilles in the Mediterranean Sea, more than five thousand miles away, where Vitol, a huge international energy trader, would take ownership of it.

The *Theo T*'s seemingly routine journey was anything but. Two weeks earlier President Barack Obama had lifted the ban that for some four decades had essentially prohibited the export of American crude oil.

Ever since a series of 1970s era laws, all of which were passed during crippling fears of oil shortages, the export ban existed as both a great rebuke and a great contradiction. On one level, the ban flew in the face of the free market ideals America holds so dear. But even as presidents from Ford to both Bushs emphasized the importance of "energy independence," the country had in fact become more and more dependent, particularly on the Middle East, and more and more embroiled in the region's politics. By the spring of 2006, U.S. net imports of crude oil and petroleum accounted for almost two-thirds of our consumption.

By the time *Theo T* set sail, carrying the first American oil export of the twenty-first century, the energy world had been entirely turned upside down by an epic development few had foreseen. America was an oil powerhouse, ready to eclipse both Saudi Arabia and Russia, and was the world's largest producer of natural gas.

Few people saw this coming. This remarkable transformation in the U.S. was brought about by American entrepreneurs who figured out how to literally force open rocks often more than a mile below the surface of the earth, to produce gas, and then oil. Those rocks—called shale, or source rock, or tight rock, and once thought to be impermeable—were opened by combining two technologies: horizontal drilling, in which the drill bit can travel well over two miles horizontally, and hydraulic fracturing, in which fluid is pumped into the earth at a high enough pressure to crack open hydrocarbon bearing rocks, while a so-called

14   proppant, usually sand, holds the rocks open a sliver of an inch
so the hydrocarbons can flow. A fracking entrepreneur likens
the process to creating hallways in an office building that has
none—and then calling a fire drill.

In November 2017, production topped the ten million
barrel a day record set in 1970, back in the last gasp of the legendary oil boom. This year, U.S. oil production is expected to
reach almost eleven million barrels a day, according to the U.S.
Energy Information Administration. The country's newest
hot spot, Texas's Permian Basin, now ranks second only to
Saudi Arabia's legendary Ghawar oil field in production per
day, according to oil company ConocoPhillips.   Stretching
through northern Appalachia, the Marcellus Shale could be
the second largest natural gas field in the world, according to
geologists at Penn State. Shale gas now accounts for over half
of total U.S. production, according to the EIA, up from almost
nothing a decade ago.

Last year, the U.S. imported less than one-third of its daily
oil demand, and the Energy Information Administration says
it's possible the U.S. will become a "net petroleum exporter,"
meaning that the amount of exports will more the offset the
amount of imports, by 2022. "It [shale] is monstrous," says
Will Fleckenstein, who drilled his first horizontal well in 1990
and is now a professor of petroleum engineering at the Colorado School of Mines. In part due to ongoing improvements in
technology, he says, "It is impossible to overstate the hydrocarbons that it is technically and economically feasible to
produce."

The apparent new era of American energy abundance has already had a profound impact around the world. Economies from Russia to Saudi Arabia to Venezuela that were dependent on the high price of oil are struggling, a situation that would have been unthinkable in a world of $100 a barrel oil, and one that is playing out in strange and unpredictable ways.

More upheaval seems inevitable as America reevaluates its strategic goals. CME Group executive director and senior economist Erik Norland calls fracking "one of the top five things reshaping geopolitics." Ever since President Franklin D. Roosevelt met the first Saudi king, Abdul-Aziz al Saud, aboard the *USS Quincy* in the Suez Canal in 1945, we've had a devil's bargain: our protection in exchange for their oil. The superficial analysis boils down to a simple question: If America doesn't need Saudi oil, does America need Saudi Arabia?

Under the Trump Administration, the longstanding dream of energy independence has taken a grander, more muscular turn. Secretary of the Interior Ryan Zinke talks about opening more federal lands like national parks to drilling in order to ensure "energy dominance." "We've got underneath us more oil than anybody, and nobody knew it until five years ago," President Trump told the press aboard Air Force One in the summer of 2017. "And I want to use it. And I don't want that taken away by the Paris Accord. I don't want them to say all of that wealth that the United States has under its feet, but that China doesn't have and that other countries don't have, we can't use."

16   Just what the United States has under its feet is in many ways
        still a mystery.

        To date, most of the complaints about fracking have
        focused on environmental concerns. (Even the term "fracking"
        is viewed by the industry as a pejorative, as it was created by
        environmentalists; in its editorial style guide, the Colorado
        School of Mines says the word "should be avoided" and sug-
        gests "fracturing" instead.) These concerns aren't the topic
        of this book, because they've been covered extensively else-
        where, because the science is still evolving—and because there
        are other, less well-known, reasons to question the notion that
        a plentiful supply of oil and gas is going to assure our future,
        extricate us from the Middle East, and allow us to crush Russia,
        OPEC, and everyone else.

        Start with the notion of energy independence. Even its big-
        gest, most informed proponents admit that what they're really
        talking about, under the most optimistic scenario, is that North
        America, including Canada and Mexico, will produce as much
        energy as it consumes. Right now, the U.S. consumes twenty
        million barrels a day, and produces just over half of that. Even
        if our production somehow catches up to our consumption,
        it won't ever completely offset the need for imports because
        of nitty gritty issues, like demand for different types of oil. "I
        got through seven years without ever saying the words 'energy
        independence,'" says a former member of the Obama Adminis-
        tration who dealt with this issue. "It's a dangerous notion." Even
        more, the market for oil is a global one, and the price of a barrel
        of oil will continue to be influenced by events outside of any-
        one's control. The oil business goes through unsettling boom

and bust cycles, and fracking does nothing to change that fact. It
is an unpredictable business.

The biggest reason to doubt the most breathless predictions about America's future as an oil and gas colossus has more
to do with Wall Street than with geopolitics or geology. The
fracking of oil, in particular, rests on a financial foundation that
is far less secure than most people realize.

The most vital ingredient in fracking isn't chemicals, but
capital, with companies relying on Wall Street's willingness to
fund them. If it weren't for historically low interest rates, it's
not clear there would even have been a fracking boom.

"You can make an argument that the Federal Reserve is
entirely responsible for the fracking boom," one private equity
titan told me. That view is echoed by Amir Azar, a fellow at
Columbia University's Center on Global Energy Policy. "The real
catalyst of the shale revolution was . . . the 2008 financial crisis
and the era of unprecedentedly low interest rates it ushered in,"
he wrote in a recent report. Another investor puts it this way:
"If companies were forced to live within the cash flow they produce, U.S. oil would not be a factor in the rest of the world, and
would have grown at a quarter to half the rate that it has."

Worries about the financial fragility of the fracking revolution have simmered for some time. John Hempton, who runs
the Australia-based hedge fund Bronte Capital, recalls having
debates with his partner as the boom was just getting going.
"The oil and gas are real," his partner would say. "Yes," Hempton
would respond, "but the economics don't work." Thus far, the
fracking industry, which survived what some saw as a con-
certed attempt led by Saudi Arabia to destroy it, has been more

18 resilient than anyone would have dreamed. But questions about
the sustainability of the boom are no longer limited to a small
set of skeptics. Those doubts now extend to the boardrooms
of some big investors as well as to the executive suites of at least
a few of the fracking companies themselves. The fracking boom
has been fueled mostly by overheated investment capital, not by
cash flow.

If that story has a central character, it's Aubrey McClendon,
the founder of Chesapeake Energy, a startup which grew into a
colossus and, for a brief moment in history, most represented
American fracking to the world. No one was more right, and more
wrong, bolder in his predictions or more spectacular in his fail-
ures, more willing to risk other people's money and his own, than
McClendon. Or as one banker who knew McClendon well puts, it,
"The world moves when people who like risk take action."

McClendon was born in 1959, when American oil still ruled
the world, a year before OPEC was created. He died in a fiery car
crash just three months after President Obama lifted the export
ban. "He was the good face of the industry—the passion, the
creativity, the daring," another former investment banker tells
me. "But he was also the bad face." And that duality makes him a
perfect personification of America's fracking revolution.

Shale Revolution

Part One

America's Most Reckless Billionaire

His death, like his legacy, was a hotly contested subject. On March 2, 2016, just after 9:00 a.m., Aubrey McClendon slammed his Chevrolet Tahoe SUV into a concrete viaduct under a bridge on Midwest Boulevard in Oklahoma City, dying instantly. He was speeding, wasn't wearing a seatbelt, and didn't appear to make any effort to avoid the collision.

Just one day earlier, a federal grand jury had indicted McClendon for violating antitrust laws during his time as the CEO of Chesapeake Energy. Investigators ultimately ruled it an accident, but rumors of suicide persist to this day. As Captain Paco Balderrama of the Oklahoma City Police told the press, "We may never know one-hundred percent what happened."

In the fall of 2008, *Forbes* had ranked McClendon number 134 on its list of the 400 richest Americans, with an estimated net worth of over $3 billion. But because he borrowed so much money, and secured business loans with personal guarantees, two years after his death, lawyers were still wrangling over the remains of his estate, trying to figure out which debts would be paid—from the $500,000 he owed the Boy Scouts of America to the $465 million he owed a group of Wall Street creditors, including Goldman Sachs. Wall Street's vultures—hedge funds that invest in distressed debt—had descended, buying the debt for less than 50 cents on the dollar, essentially rendering a judgment that the claims wouldn't be paid in full.

If McClendon did die broke, it wouldn't have been out of character. During his years as an oil and gas tycoon, he fed on risk, and was as fearless as he was reckless. He built an empire that at one point produced more gas than any American company except ExxonMobil. Once, when an investor asked on a conference call, "When is enough?" McClendon answered bluntly: "I can't get enough."

Many think that without McClendon's salesmanship and his astonishing ability to woo investors, the world would be a far different place today. Stories abound about how at industry conferences, executives from oil majors like Exxon would find themselves speaking to mostly empty seats, while people literally fought for space in the room where McClendon was holding forth.

"In retrospect, it was kind of like Camelot," says Henry Hood, Chesapeake's former general counsel, who worked at Chesapeake initially as a consultant from 1993 until the spring of 2013. "There was a period of time that will never be duplicated with a company that will never be duplicated."

"Aubrey was a very curious person, and that single trait led him to succeed," says Marc Rowland, who got to know McClendon in the early 1980s and served as Cheseapeake's CFO from 1992

22 to 2010. "A lot of people are driven and smart, but they lack curiosity. Aubrey had that in spades."

Many people have a far less favorable opinion of McClendon. "Aubrey is irrelevant," one oil executive tells me. "If you want to tell the American success story, you'll ignore him. If you want to tell the sad story, write about him."

Some of his peers, along with some on Wall Street, considered him a buffoon, a con man of sorts, and maybe even a fraud. "He was a catalyst and a visionary, sure, but he tried to kiss all the girls," says one old-time oil man. "He was a whirling dervish." "America's Most Reckless Billionaire," *Forbes* once called McClendon, and for many in the industry, that headline defined the man.

But if it was a con, he was conning himself, too. Because he *believed.*

He was, in many ways, the embodiment of a transformation that has changed the face of not just the oil and gas industries but of geopolitics as well. The contradictions and questions in McClendon's story continue to reverberate across the industry he did so much to create. You might think of McClendon as a bit of J. R. Ewing, the fictional character in the television series *Dallas,* mixed with Michael Milken, the junk bond king who pioneered an industry and arguably changed the world, but spent several years in prison after pleading guilty to securities fraud. Over and over, I heard the same refrain: "Aubrey epitomizes everything we're talking about."

Unlike many others who come from nothing and make their fortunes in the oil patch, McClendon, who was born on July 14, 1959 in Oklahoma City, was oil industry royalty. His great uncle was the Oklahoma governor and senator who in 1929 co-founded Kerr-McGee, which was the ExxonMobil of its time. McClendon, who was always immensely popular, was the president and co-valedictorian of his senior class. He headed to Duke, where he was the rush chairman of his fraternity. "Athletes and non-athletes, party boys and geniuses," is how he described those years to a Duke University publication. "It was a collection of good guys from across the nation. We studied hard, we played hard." "He was super competitive and aggressive," recalls someone who knew him at Duke. "If he had a few drinks, he'd want to wrestle. He was big and strong and a little bit out of control."

At Duke, McClendon met the woman who would become his wife, Whirlpool heiress Kathleen Upton Byrns. Her cousin, Fred Upton, has served as a Republican congressman from Michigan since 1987. As chairman of the Committee on Energy and Commerce, Upton was a key defender of fracking.

In college, he also met Hood, who recalls a driven if rambunctious young man. "Aubrey was thoughtful, tall, and handsome, but incredibly clumsy," Hood recalls. "He always had an ink stain on his shirt from the pen in his front pocket. We called him 'Aubspill,' because he was always spilling. In basketball, he was always throwing elbows like a bull in a china shop." Another variation of the nickname, Hood recalls, was " 'Aubkill,' because McClendon's outsized competitive instinct made him dangerous in physical activities."

24   McClendon thought about being an accountant until, during his senior year, he came across an article in the *Wall Street Journal*. "It was about two guys who had drilled a big well in the Anadarko Basin that had blown out, and it was alleged to be the biggest blowout in the history of the country," McClendon told *Rolling Stone*. "They sold their stake to Washington Gas and Light and got a $100 million check. I thought, 'These are two dudes who just drilled a well and it happened to hit.' So that really piqued my interest."

Three years before McClendon was born, the iconoclast geologist M. King Hubbert first outlined his ideas on peak oil. Essentially, the idea was that since the amount of oil is finite, production will follow a bell curve, and after peaking, it will inevitably decrease. Time seemed to prove Hubbert right. American oil production peaked in 1970 at 9.6 million barrels a day and began a steady, seemingly inexorable decline.

For the U.S., it was a profound role reversal. Until the 1970s, American oil and swaggering Texas oil barons ruled the world. The Texas Railroad Commission controlled the international price of oil by allowing a certain amount of production, and maintaining spare capacity. But in 1972, as U.S. production slowed, Texas had to start producing flat out. "This is a damn historic occasion and a sad occasion," the Texas Railroad Commission's chairman declared. The following year OPEC, which had been created by Iran, Iraq, Saudi Arabia, and Venezuela in 1960, began to flex its new muscles. OPEC declared an oil embargo during the Yom Kippur war against all of Israel's allies, including the United States. Oil prices quadrupled.

The disruption of oil sent Americans in search of other energy sources. Yet oil, which is primarily used for transportation (passenger cars today account for about half our daily consumption), has the benefit of being relatively easy to ship around the world. Natural gas, on the other hand, is used primarily for heating and in power plants and manufacturing. It cannot easily be shipped. One must construct liquefaction plants in order to freeze the gas into liquid form, and then build "regasification" plants to turn the LNG back into gas, all of which costs billions of dollars. And in the 1970s, the scarcity of gas was a major concern too. Congress effectively kept the U.S. from building gas-fired power plants in favor of coal in 1978. The government also launched a partnership between the Energy Department and dozens of companies and universities called Eastern Gas Shales Project, which aimed to figure out how to recover natural gas from shale deposits. And in 1980, Congress passed the Crude Oil Windfall Tax Credit Act, which provided tax credits to "qualified unconventional gas wells." By "unconventional," they meant drilling in areas, like the shale rock, that had not been drilled before.

When McClendon graduated from Duke in 1981, everyone said oil prices would only go higher. But prices defied the prognosticators and began to crater, thanks to a global economic recession and a tidal wave of new supply from United Kingdom's North Sea, Alaska's North Slope, and Mexico. OPEC cut its production in an effort to boost the price, but with the onslaught of new supplies, all that happened was that Saudi Arabia lost market share. In 1985, Saudi Arabia gave up and unleashed production, and the price tumbled further.

26   Not incidentally, that was also the last gasp of truly spectacular American oil riches. In the early 1980s, drilling in Texas,
in particular, had rebounded as prices shot higher. It was the era
of J. R. Ewing-style conspicuous consumption. Midland, Texas
even boasted its own Rolls Royce dealership. But as the price of
a barrel of crude slid, real estate cratered and banks went under.
Occidental Petroleum bought Iowa Beef Processors, and Gulf
Oil considered buying the Barnum and Bailey Circus. Oil had
become a grungy, dreary business. It was desperate days.

But McClendon was never one to be deterred. He thought
there was opportunity in assembling packages of drilling
rights—for gas, not oil—either to be sold to bigger companies or to be drilled. In the mere existence of that opportunity,
America is almost unique, because it is one of the few countries
where private citizens, rather than governments, own the min-
eral rights under their properties. In order to drill, you just have
to persuade someone to give you a lease. McClendon became
what's known in the oil and gas business as a land man, the
person who negotiates the leases that allow for drilling.

That, it turned out, would make him the perfect person for
the new world of fracking, which is not so much about finding
the single gusher as it is about assembling the rights to drill
multiple wells. "Landmen were always the stepchild of the
industry," he later told *Rolling Stone*. "Geologists and engineers
were the important guys—but it dawned on me pretty early that
all their fancy ideas aren't worth very much if we don't have a
lease. If you've got the lease and I don't, you win."

In 1983, when McClendon was just twenty-four years old,
he partnered with another Oklahoman named Tom Ward,

"doing deals for scraps of land in Oklahoma, faxing each other in
the middle of the night," Ward later told *Rolling Stone*. Six years
later, the two thirty-somethings formed Chesapeake Energy,
which was named after the beloved bay where McClendon's
family vacationed. They seeded it with a $50,000 investment.

In some ways, they were an odd couple. Bespectacled and
balding, Ward came across as more of a typical businessman,
whereas McClendon, with his flowing hair and Hollywood good
looks, was the dynamo of the duo. They divided the responsi-
bilities, with McClendon happily playing the front man, raising
money, and talking to the markets, while Ward stayed in the
background running the business. They operated out of separate buildings, with separate staffs. Much later, one observer
recalls that when the two would go to Oklahoma City Thunder
basketball games, their blocks of seats were on opposite sides of
the arena, and they never sat together.

Neither Ward nor McClendon were technological pioneers.
That distinction, most people agree, goes to a man named
George Mitchell, who drew on research done by the government to experiment on the Barnett Shale, an area of tight rock
in the Fort Worth basin of North Texas. Using a combination
of horizontal drilling and hydraulic fracturing, Mitchell's team
cracked the code for getting gas out of rock that was thought to
be impermeable.

The few people who were paying attention to what Mitchell
was doing were far from convinced that it would succeed. Giants
like Exxon were selling off their U.S. properties to the small
independent companies and going international. "At the time,
we dismissed shale because ExxonMobil told us it would cost

28 $125 a barrel to get it out and would never work," says Jeff Currie, the global head of commodities research at Goldman Sachs.

McClendon, however, was the pioneer in the other essential part of the business: raising money. "As oxygen is to life, capital is to the oil and gas business," says Andrew Wilmot, a Dallas-based mergers and acquisitions adviser to the oil and gas industry at Purposed Ventures. "This industry needs capital to fire on all cylinders, and the founder and father of raising capital for shale in the U.S. is Aubrey McClendon." "To be able to borrow money for ten years and ride out boom and bust cycles was almost as important an insight as horizontal drilling," McClendon, with typical immodesty, later told *Rolling Stone*. "I never let Aubrey McClendon in the door for a meeting," says an analyst who works for a big investment firm. "Because we would have bought a ton of stock and it would not have ended well. He was that good."

In the early 1990s, Bear Stearns helped Chesapeake sell high-yield debt in a first-of-its-kind sort of deal. This was no small achievement. After all, Chesapeake didn't have much of a track record, and there was less than zero interest in the oil and gas business from the investment community. "I watched him convince people in these meetings," says a banker who was there. "He was so good, so sharp, with such an ability to draw people in."

On February 12, 1993—a day McClendon would later describe as the best one of his career—he and Ward took Chesapeake public. They did so despite the fact that their accounting firm, Arthur Andersen, had issued a "going concern" warning, meaning its bean-counters worried that Chesapeake might go out of business. So McClendon and Ward simply switched accounting firms. "Tom and I were thirty-three-year-old land men at the time, and most people didn't think we had a clue of what we were doing, and probably in hindsight they were at least partially right," McClendon told one interviewer in 2006. The IPO reduced their ownership stake to 60 percent, but both men kept for themselves an important perk, one that would play a key role in the Chesapeake story: They got the right to take a personal 2.5 percent stake in every well Chesapeake drilled. In the years following its IPO, Chesapeake was one of the best-performing stock on Wall Street, climbing from $0.47 per share to $34.44 per share.

The story that drew in investors was set in a place called the Austin Chalk, which McClendon made sound almost magical— never mind that the *Texas Monthly* had once called it the "most perverse, contrary, incorrigible oil field known to man." Its limestone straddles the border between Texas and Louisiana, and while everyone knew that oil was there, the rock wasn't porous enough to get it out.

Then in 1994, a company called Occidental drilled a hugely successful horizontal well there. Mitchell hadn't yet done his pioneering work, but Occidental showed that horizontal drilling could allow companies to extract vast quantities of gas economically in a way that hadn't previously been possible.

McClendon went all in. Chesapeake leased more than a million acres of the Austin Chalk, and McClendon told the *Oil & Gas Journal* that the location could be "the largest onshore play in the country." He projected that Chesapeake's production of gas would grow by 50 percent a year. As the stock soared

30   and Chesapeake issued ever more optimistic press releases, Chesapeake sold approximately $1 billion worth of equity and debt, according to a lawsuit that was later filed. On April 2, 1997, a press release announced Chesapeake's completion of a well called the 1-H Brown, which the firm said was "the most productive horizontal well ever drilled and the most productive well of any type drilled onshore in the U.S. during the past ten years." A young stock analyst named John Raymond, who worked at Howard, Weil, Labouisse, Friedrichs, Inc., was also bullish on the Chalk, in large part because he was a big believer in the ways new drilling technologies would reshape the industry. (Raymond's father was Lee Raymond, who was then the CEO of Exxon.) In the years following its IPO, Chesapeake was one of the best-performing stocks on Wall Street, climbing from $1.33 a per share (split adjusted) to almost $27 per share.

The grand proclamations drew the attention of short sellers. Short sellers are essentially the skeptics of Wall Street. Instead of trying to find good stocks, they try to find bad ones, and they make their money when the stock declines. Among other things, so-called "shorts" argued that while there might be sweet spots within the Chalk where a well would produce stupendous results, that didn't mean the entire area, or even most of it, would produce the same results.

The skeptics were right. In the spring of 1997, Chesapeake announced that much of the land it had acquired was not productive. The company took a $200 million charge against earnings, which essentially wiped out all the profit it had declared in the previous three years, and the stock plunged 25 percent.

The loss coincided with the Asian financial crisis, which sent oil and natural gas prices plummeting. By 1998, Chesapeake was selling for seventy-five cents a share. McClendon and Ward tried to sell the company, but there were no takers. Later, McClendon told *Oil & Gas Investor*: "To look at the quote machine screen every day back then and think, 'You're not even worth one dollar' was probably the worst period of our careers."

As he would do again and again, McClendon survived by borrowing yet more money to acquire more properties. "Simply put, low prices cure low prices as consumers are motivated to consume more and producers are compelled to produce less," he wrote in Chesapeake's 1998 annual report. McClendon essentially made a giant bet that gas prices would rise on their own, and he billed the properties he acquired "low risk." Luckily, he was right. Within a few years, prices were soaring again, and McClendon had gotten out of a jam, for now.

In 2001, George Mitchell became one of the first of many fracking billionaires, when his company was bought by Devon Energy. Others began to notice that the production from Devon's wells in the Barnett Shale was phenomenal, and a paper presented at the American Association of Petroleum Geologists meeting in Dallas argued that the Barnett formation possessed two and a half times the amount of gas that they'd previously estimated.

By accident, Chesapeake held leases in the Barnett too. In 2002, Chesapeake bought a company called Canaan Energy. The deal included 7,000 acres in the Barnett as a "sweetener" used to justify the $118 million price tag, as detailed in Gregory Zuckerman's book *The Frackers: The Outrageous Story of the New*

32   *Billionaire Wildcatters*. After seeing Devon's filings, Chesapeake began to snap up as many acres as it could in the Barnett.

In the ensuing years, the rush for what became known as "shale plays" exploded beyond the Barnett. There was also the Fayetteville, the Haynesville, the Marcellus, and more. When a study projected that the impact of the Fayetteville shale play in Arkansas would total about $18 billion over the five years, McClendon told a luncheon crowd at the Arkansas Economic Development Foundation, "We're about to make that completely irrelevant. The way I see it, [natural gas companies] are going to spend somewhere between $75 [billion] to $100 billion in your state over the next decade or so."

By the end of 2004, Chesapeake had spent around $6 billion over the previous decade acquiring properties, companies and leases. Over the next four years, the company spent another $21.5 billion. On Wall Street, Chesapeake was much beloved because of the fees the company paid the banks to raise all the money it needed to fund itself. From 2001 to 2012, Chesapeake sold $16.4 billion of stock and $15.5 billion of debt, and paid Wall Street more than $1.1 billion in fees, according to Thomson Reuters Deals Intelligence. McClendon was like no other client. "Aubrey would send an email to six or seven banks saying, 'Hey, here's what I need. Come back to me with your best offer,'" one investment banker says. When one particular deal closed, he received a box from McClendon with bottles of Cristal and a nice note, as did everyone else who was involved.

McClendon, who would later call these years the "The Great North American Land Grab," developed a reputation among his peers for overpaying. Stories abounded of him offering ten times the amount of other offers. "Last night, I got back to my room at about 2:00 a.m.," he told a gathering of oil men in Houston in 2005. "I went through some emails, and there's no telling what I did. So if I bought you, I probably overpaid. Congratulations." "You wouldn't see Aubrey out late at night," says Rob Lambert, a portfolio manager at Nassau Re. "He'd be emailing you from his hotel room at 4:00 a.m. He lived a big life, but he was an extremely hard-working family man,"

"He could buy acreage faster than I could fund it," recalls Rowland, a fact that eventually led to his resignation as CFO in 2010 (with McClendon's blessing).

His aggressiveness didn't endear him to the old-time oil men. "Everyone in Midland hated Chesapeake," one says. "They came out here when land was leasing for $200 to $300 an acre. All of a sudden, Chesapeake was paying $2,000 to $3,000 an acre. They got in some good places because they shut everyone else out. Their attitude was, 'We are Chesapeake, get out of our way.'" "His aggressive style ruffled some feathers in the industry," Andrew Wilmot says. "He went after new plays guns blazing, and drove up the prices. That made some people millionaires, but it wreaked havoc on others."

But with the sort of price increase that the market was experiencing at the time, it didn't seem to matter what McClendon had paid. Gas prices steadily marched upward, and by their peak in June 2008, they had more than doubled in just a few years. Chesapeake's stock moved in lockstep, recovering its losses from the 1990s, and more. It hit over $65 a share in the summer of 2008, giving the company a market value of more than $35 billion. That made McClendon's shares worth some $2 billion.

34   McClendon went on a corporate spending spree that would have put today's Silicon Valley chieftains to shame. "Asking me what to do with extra cash is like asking a fraternity boy what to do with the beer," McClendon told *Natural Gas Intelligence* in 2005. Chesapeake's campus in Oklahoma City boasted a 63,000-square-foot daycare center with room for 250 children, a luxurious gym, and multiple cafés with actual chefs. Nor was McClendon frugal when it came to his personal life. He acquired multimillion dollar mansions and resorts in Oklahoma, Bermuda, Maui, Vail, on Lake Michigan, and even in Minnesota. He had one of the best wine collections in the world.

In the summer of 2006, McClendon, along with other partners (including Chesapeake's Tom Ward, who had retired from the company earlier that year, later explaining to Zuckerman "all day, then nights and weekends, it didn't stop") bought the Seattle SuperSonics and the Seattle Storm basketball teams for about $350 million from Starbucks CEO and funder Howard Schultz. In 2007, McClendon was hit with a $250,000 fine from the NBA for telling a reporter that "we didn't buy the team to keep it in Seattle," which was contrary to what the NBA had been led to think. Indeed, the next year, the purchasers moved the SuperSonics to Oklahoma City and renamed them the Thunder. The public anger was such that Schultz even sued to rescind the sale.

To Wall Street investors, McClendon was delivering on what they wanted most: consistency and growth. His pitch was that fracking had transformed the production of gas from a hit or miss proposition to one that operated with an on and off switch.

It was manufacturing, not wildcatting. He became a flag waver for natural gas—"Mr. Gas," *Fortune* magazine once called him. "Aubrey was the first one to say, 'Let's create demand,'" Henry Hood says.

Back in 2003, when McClendon was just getting started, the consensus view had been that the U.S. was running out of natural gas. It became a fixation for Alan Greenspan, the once-revered chair of the Federal Reserve, who warned Congress during a rare appearance that the shortage and rising cost of gas could hurt the American economy. Greenspan recommended that the U.S. build terminals to accept deliveries of LNG from other countries. "We see a storm brewing on the horizon," said Representative Billy Tauzin, Republican of Louisiana and the then-chairman of the Energy and Commerce Committee. (Such fears eventually helped push through the Energy Policy Act of 2005, which exempted natural gas drillers from having to disclose the chemicals used in hydraulic fracturing, thus averting costly regulatory oversight.)

As fracking took off, McClendon began telling anyone who listened that the U.S. had enough natural gas to last more than a hundred years. He quietly financed a campaign called "Coal is Filthy," and he argued that converting 10 percent of U.S. vehicles to natural gas in the next ten years would be the fastest, cheapest way to free the country from dependency on foreign oil. He was adamant that employees drive cars fueled by compressed natural gas.

For a man steeped in the industry's history of booms and busts, McClendon had by now convinced himself that gas prices would never fall. In August 2008, he predicted that gas would

36   stay in the $8 to $9 range for the foreseeable future. "He had a very, very strong point of view about gas," says a banker who knew him since the early 1990s. "By the way, he was basically wrong for the last thirty years."

But as was almost always the case, if McClendon believed something, he'd seldom have difficulty getting others to buy in. That spring and summer, Chesapeake raised another $2.5 billion by selling stock, and another $2 billion selling debt. McClendon had bought 750,000 more shares that were simply borrowed against stocks he already held, he later told *Forbes*. "He was always aggressive in his point of view, always aggressive with leverage, always willing to be all in," this banker says. "He was going to bet the farm, and if he lost, he was going to bet the farm again. Most of us, if we get the farm, we don't want to lose it! He didn't have a regulator. It's a personality type. So sure you're right, no boundaries, no ability to say, not this one. He had to win. It was so important to him to win."

McClendon's bullish view on prices became the conventional wisdom in energy markets. In 2007, the National Petroleum Council pronounced that domestic natural gas supplies would be insufficient to satisfy demand, and Congress passed the Energy Independence and Security Act of 2007, which, predictably, "aimed to move the U.S. toward greater energy independence," this time by increasing the production of renewables. That year, the supposedly smartest investors in the world—among them Goldman Sachs and takeover titan KKR—structured their massive $45 billion buyout of a utility called TXU in a way that was essentially a bet that natural gas was going to be worth much higher than the $7 price tag around then. Plans to build dozens of multibillion dollar facilities to import LNG had been announced.

At the same time, Vladimir Putin was making similar bets. In an attempt to set up an OPEC-like cartel for gas, the Russian premier hosted a group of gas-producing countries, including Algeria, Iran, and Venezuela, in Moscow. The U.S. was not among them. "Costs of exploration, gas production, and transportation are going up," Putin said. "It means the industry's development costs will skyrocket. The time of cheap energy resources, cheap gas, is surely coming to an end."

But if McClendon was right that lower prices cure lower prices, he somehow forgot the flip side of that industry truism—that high prices also cure high prices. Time and again, in commodity markets, high prices encourage more producers to produce, creating a surplus, that then crushes prices—and producers. "He was right that shale changed the world," says a longtime gas man. "He should have listened to himself."

The Brain Trust

One fracking enterprise, whose pedigree might alarm ordinary investors, saw the difficulties ahead. This company maintained a far lower profile than McClendon's Chesapeake, but its trajectory may tell the other side of the fracking story—one based on technological innovations and sound financial footing.

In 1999, when McClendon was struggling to recover from the Austin Chalk debacle, a Texas energy powerhouse called Enron, whose CEO Jeff Skilling had become dismissive of any businesses that required hard assets, spun off an oil and gas exploration division called, appropriately enough, Enron Oil and Gas. The company promptly renamed itself EOG Resources. In those days of the first Internet boom, no one much cared, and the stock languished. Then, EOG began applying horizontal drilling techniques in the Barnett Shale. As others caught on and EOG's gas production began to soar, the stock began to run up. "That turned around the company, and focused us as a shale company," says current CEO William "Bill" Thomas. "We were first movers in the shale revolution." Today, EOG is valued at almost $70 billion, more than Enron was at its peak before its infamous collapse.

If EOG was the anti-Enron, it was also the anti-Chesapeake. Instead of a splashy campus, EOG, which prided itself on a decentralized culture, operated out of several floors in a nondescript office tower in Houston. Among investors, EOG became known for technological advances. "Everyone else is lucky," one old-time oil and gas man says. "EOG is good." Others call EOG "the Harvard of Shale," "the Apple of Oil," or simply, "the Brain Trust."

In the mid to late 2000s, most of EOG's revenue came from producing gas. But Mark Papa, a former petroleum engineer who was the CEO of the company from 1999 until he handed the reins to Thomas in late 2013, realized that natural gas prices would be low for several decades. They needed to become an oil company "or we're dead ducks," he told management in 2007, according to Zuckerman's book.

"I did my own macro homework and it was glaringly obvious to me that with EOG and others finding such huge supplies of natural gas, and with zero export capability, there was going to be huge oversupply," Papa says now. "Collapse was inevitable. So as a corporate strategy we had to literally run away from North American natural gas—and we were a North American natural gas company!" He adds, "It is incredible to me that others didn't see the train coming down the tracks."

In oil, first came the Bakken. Geologists had known since the 1950s that a formation of about 200,000 square miles

40   below parts of Montana, North Dakota, and Saskatchewan, called the Bakken, contained oil. The Bakken isn't technically considered shale, but rather siltstone.

EOG—and an independent energy producer named Continental Resources, run by Harold Hamm (rumored to be Trump's pick for the Secretary of Energy before Rick Perry got the job), took the lead in the Bakken. Over the next five years, oil production in the formation grew by more than tenfold to almost a million barrels a day. In 2000, North Dakota was ranked ninth among U.S. oil-producing states, and forty-third in terms of economic output per person. By 2012, it was the second-largest energy-producing state in the nation, and a one-bedroom apartment in Williston, North Dakota, rented for $2,000 a month, according to notes taken at a meeting of the North Dakota Sheriffs and Deputies Association. Farmers were charging campers $800 a month per camper; the McDonalds in Williston announced it would pay $15 an hour, plus an immediate $500 sign on bonus; the Williston General Motors dealership had become the number one seller of Corvettes in the upper Midwest; instead of stocking shelves, the Wal-Mart simply brought out pallets of merchandise and set them in the aisles; the county jail increased its bookings by 150 percent.

Even as the Bakken began to explode, the accepted wisdom still was that shale was different, and that oil molecules were too large to flow through shale swiftly enough to make fracking for oil economic. But Thomas and his team thought differently, and they began to try in the Barnett Shale, near where George Mitchell had had the first successes with gas. "Papa trusted us on the technology," says Thomas, who came up through the

geology ranks, as many do at EOG. "The industry and the aca-
demic world said, 'Never.' We were berated. Anyone who had
ever heard of it thought we were nuts." But by 2007, wells in the
Barnett were producing oil. The wells weren't spectacular, but
they were a technical success.

Thomas and his team also believed that an area called the
Eagle Ford Shale—a substrata of the Austin Chalk—might
produce oil. EOG quietly began assembling land for less than
$500 an acre. In the spring of 2010, EOG announced to a crowd
of Wall Street investors at the Houston Four Seasons that the
Eagle Ford contained over nine hundred millions barrels of oil,
enough to rival the Bakken. Zuckerman reported that people
raced out of the room to trade; within a few days, EOG's stock
hit over $100 a share, up over tenfold from the neglected years
following the Enron spinoff.

Land in the Eagle Ford that had leased for roughly $500 an
acre jumped to $5,000 an acre within the space of a year. By
early 2013, EOG was completing so-called "monster" wells that
produced over 2,500 barrels a day.

Money poured into the American energy business—which,
in the lean years following the Great Recession, was one of the
few areas to show the growth investors craved. Oil prices were
soaring, and no one thought they would ever fall again. Before
long, the boom began to reshape the U.S. economy. Between
2011 and 2014, the Wisconsin Department of Natural Resources
handed out more than a hundred permits for sand mines.
Frackers had discovered that great quantities of "Wisconsin
white" seemed to work best as proppant. From 2009 to 2012, the
amount of sand shipped by railroads more than doubled, mainly

due to frac sand, according to the American Association of Railroads. Rail giant Union Pacific even started a program called "Sand 2 Shale" to expedite the shipments of sand.

In 2007, Ed Rendell, then the governor of Pennsylvania, said the state received 71 requests for drilling permits in the Marcellus. By 2010, there were more than 3,000 requests. Demand for metals to make drills and other fracking machinery shot up. In March 2011, Nucor, a big steelmaker, broke ground on a new $750 million iron plant in Louisiana. "We could change the entire manufacturing base in the U.S. if we just embrace what's happening in natural gas," Nucor's then CEO, Dan DiMicco, told the *Wall Street Journal*.

Fortunes were made. While the media focused its gaze on the high priests of technology like Amazon's Jeff Bezos and Facebook's Mark Zuckerberg, unknown tycoons quietly raked in billions.

A man named Terrence Pegula, who was born into a coal mining family in Pennsylvania and who majored in petroleum engineering on a scholarship, ran a struggling small time drilling operation called East Resources, which he'd started by borrowing $7,500 from family and friends. It just happened to sit atop the Marcellus shale. In 2009, the firm was sold to Royal Dutch Shell for $4.7 billion. Pegula then bought the Buffalo Sabres for $189 million—and then, in 2014 outbid groups led by Donald Trump and Jon Bon Jovi to buy the Buffalo Bills as well. He also donated $90 million to his alma mater, Penn State, to help establish a Division I hockey program at the school.

In 1989, a geologist named Jeff Hildebrand founded Hilcorp Resources, just three years after he'd earned his degree. In 2011, Marathon Oil bought the company's 100,000 acres in the Eagle Ford for $3.5 billion. Hildebrand, who still runs Hilcorp (and, according to *Forbes* is worth over $4 billion), now owns the 1,000-acre ranch in Aspen that used to belong to singer John Denver. He also funded a $32 million equestrian center at Texas A&M University. In 2015, the *Houston Chronicle* reported that Hilcorp had paid every single one of its employees a $100,000 bonus.

Bob Simpson, who ran a natural gas producer called XTO Energy, was among those who saw the flaws in McClendon's strategy. In 2010, ExxonMobil, then run by Rex Tillerson, bought XTO for $35 billion, and that year Simpson and his partner paid nearly $600 million for the Texas Rangers baseball team.

In West Texas, brothers Dan and Farris Wilks wanted someone to frack a well on their ranch, but no one was interested. So they built a pump and did it themselves. That pump grew into a company called Frac Tech, which the brothers sold in 2011 for $3.5 billion. They bought Montana's 62,000-acre N Bar Ranch for $45 million. In the run up to the 2016 election, they and their wives also cut a $15 million check to a PAC backing Ted Cruz.

Also in 2011, KKR—the buyout firm whose hostile takeover of RJR Nabisco was immortalized in the book *Barbarians at the Gate*—bought a Tulsa, Oklahoma-based natural gas company called Samson Resources for $7.2 billion, including more than $4 billion in debt. Samson was run by a Yale graduate named Stacy Schusterman, who had taken over the business named for her grandfather when her father, the founder, died in 2000.

44   Serial Silicon Valley entrepreneurs have nothing on a Houston oilman named Floyd Wilson, who had sold one of his earlier companies to Chesapeake, and then another to a large company called Plains Exploration. He then started afresh with PetroHawk, which grew into one of America's largest gas producers. In 2011, Australian mining company BHP Billiton bought Petrohawk for $12 billion. Wilson promptly started yet another company called Halcon.

Another Houston oilman named Jim Flores had helped to build Plains Exploration, which tried but ultimately failed to strike a deal to frack in California's Santa Barbara County. In 2013, Plains was sold for roughly $9 billion, giving Flores a rumored payout of over $150 million. The Flores family owns a 30,000-square-foot mansion in River Oaks, Houston's most prestigious neighborhood, which was previously owned by two other famous Texas oil dynasties, the Cullens and the Wyattas. The house features a Steuben crystal staircase, and its formal gardens contain a tapestry rose garden and a camellia allee, according to photos taken on a local home and garden tour by a blogwriter called Lanabird.

Debt

If the late 2000s were the glory days, no one told Aubrey McClendon.

As gas prices began to fall in 2008, so did Chesapeake's stock, from its peak of $70 in the summer of 2008 to $16 by October. With that steep slide, the value of the shares he'd pledged to banks in exchange for loans also fell—and the banks called his margin loans. Rowland, whose office was about fifty feet away from McClendon's, recalls him walking in and saying, "Marc, they're selling me out." "It was a one minute conversation," says Rowland. "He went from a $2 billion net worth to a negative $500 million. There wasn't any sweat in his eye or anything like that. It was just the way he was."

Indeed, from October 8 to October 10, McClendon had to sell 94 percent of his Chesapeake stock. "I would not have wished the past month on my worst enemy," he said in a meeting.

Despite Chesapeake's abysmal stock performance in 2008, which destroyed some $30 billion in shareholder money, his

46   board of directors—consisting of mostly longtime friends of McClendon's and former politicians like Don Nickles, who was a Republican senator from Oklahoma from 1981 to 2005—gave him a $75 million "bonus" to bring his total pay that year to $112 million, making him the highest paid CEO in corporate America that year. Which made sense in that the Chesapeake board itself was one of the highest paid in the industry. From 2009 to 2011, Chesapeake paid $13.3 million in total compensation to ten board members who weren't Chesapeake executives, according to Reuters. By comparison, the highly profitable Exxon Mobil paid ten non-executive board members just $9.9 million over the same period. And Chesapeake board members also were allowed personal use of Chesapeake planes. "I have never seen a more shameful document than the Chesapeake proxy statement," Jeffrey Bronchick, a longtime portfolio manager who runs a firm called Cove Capital, wrote in a letter to Chesapeake's board. "If I could reduce it to one page, I would frame and hang it on your office wall as a near perfect illustration of the complete collapse of appropriate corporate governance."

Underlying all of McClendon's enterprises was a vast and tangled web of debt. That 2.5 percent stake in the profits from Chesapeake's wells that McClendon and Ward had kept for themselves at the IPO had come with a hitch: McClendon had to pay his share of the costs to drill the wells. Over time, according to a series of investigative pieces done by Reuters, he quietly borrowed over $1.5 billion from various banks and private equity firms, using the well interests as collateral. Reuters, which entitled one piece "The Lavish and Leveraged Life of Aubrey McClendon," also reported that much of

what McClendon owned, from his stake in the Oklahoma City
Thunder to his wine collection to his venture capital and hedge
fund investments, was also mortgaged.

McClendon's alter ego was in a similar position. Chesapeake had ignored the Bakken because McClendon hadn't seen
the potential in fracking for oil. As gas prices fell, the company
scrambled to reshape itself by building a position in the Eagle
Ford and in other hot oil plays. The payoff that was supposed to
come from the years of investing didn't arrive. Overall, Chesapeake bled cash. From 2002 to the end of 2012, there was never
a year in which Chesapeake reported positive free cash flow
(meaning the cash it generated from operations less its capital expenditures.) Over the decade ending in 2012, Chesapeake
burned through almost $30 billion.

To bridge the gap between what the business produced and
its costs, Chesapeake loaded up with debt and sold stock to
investors, and its frantic fundraising by no means ended there.
In partnership with a fraternity brother from college named
Ralph Eads, who had gone into investment banking, McClendon
scoured the globe for investors who wanted what he was selling.
From 2008 to 2012, Chesapeake cut deals with a number of
global investors, from the China National Offshore Oil Corpora-
tion to India's Reliance Energy to Australia's BHP Billiton. The
buyers would acquire some portion of Chesapeake's acreage in
the latest hot shale play, and sometimes agree to cover both par-
ties' drilling costs.

The *New York Times* calculated that from 2008 to 2012,
Eads's firm, Jefferies & Co., helped Chesapeake raised $33.7
billion. He told investors that the American shale revolution

was an opportunity they simply could not afford to pass by. "This is like owning the Empire State Building," Eads told the *Times*. "It's not going to be repeated. You miss the boat, you miss the boat."

If you include the cash Chesapeake made on its sales of acreage, its cash flow picture looks much better, albeit still negative—but Chesapeake was supposed to be in the business of producing energy, not the business of flipping land.

In 2012, McClendon went to Asia and had fifty-two meetings with investors from New Delhi to Seoul in the course of two weeks. "We have the assets they want, and we need their money," he told Bloomberg. "McClendon believes by the end of next year, the big exploratory and land grab (in the shales) will be done, and there will be no more big oil or gas basins to be discovered," an investment analyst wrote in a brief. "It worked as long as there was a next play," says one industry skeptic. "If you run out of investors to sell it to or people stop believing the story, then you have a problem." By spring, credit rating agency Moody's reported that while Chesapeake had $12 billion in debt on its balance sheet, deals like this had raised the real total to $23.6 billion.

Broke or rich, there wasn't much of a change in McClendon's demeanor. There were more new shale plays, from the Utica—which McClendon claimed would generate over half a trillion dollars in revenue for Ohio and would be "biggest thing to hit the state of Ohio economically since maybe the plow"—to the Cline shale in Texas, where local leaders in the town of Sweetwater, atop the shale, spent tens of millions to upgrade the county courthouse and the hospital in anticipation of the boom to come, according to the *Fort Worth Star Telegram*. "Shale oil

field 3 times bigger than the Eagle Ford (and 6x bigger than the
Bakken)," read one headline.

As for the crushingly low gas prices, McClendon argued
that this simply created a different sort of opportunity. Gas at
$2, he said, was providing "an $800 million daily boost to the
economy." He argued that Americans could save money, the
environment, and our country by converting transportation to
natural gas.

But the reality was unchanged: McClendon needed higher
prices. And he thought he was going to get them. Among other
things, he struck a deal with a pipeline company to transport
Chesapeake's natural gas, the terms of which quickly became
onerous if Chesapeake didn't produce ever-growing quantities of
gas, regardless of prices. "He designed the company in such a way
that it would really suffer in a downturn," says one close observer.
"He became the poster boy for a bad shale business model."

In the spring of 2012, the series of investigative stories
by the Reuters reporting team revealed the existence of the
billion-dollar-plus loans McClendon had taken out to cover his
portion of the drilling costs on Chesapeake's wells. It would later
turn out that McClendon had also backed some of the loans with
personal guarantees. The bulk of the loans came from the private
equity firm EIG Global Energy Partners. EIG was simultaneously
helping Chesapeake itself finance the purchase of assets, raising
concerns that McClendon had a huge conflict of interest. Investment analysts told Reuters they'd had no idea about the loans.
Over the next month, Chesapeake's stock fell 30 percent.

In the spring of 2012, Chesapeake's board stripped
McClendon of his chairmanship, and over the course of the next

50 year, new investors who thought they could clean up Chesapeake, including Carl Icahn, took seats on the Chesapeake board.

Icahn, according to someone familiar with events, initially thought he could control McClendon's most reckless impulses, especially with McClendon no longer serving as both CEO and chairman, but began to realize it was futile. Lou Simpson, who had run Geico's investment portfolio for Warren Buffett for decades, and who had joined the board in 2011, was also furious at the state of affairs, according to someone close to events. In remarks in front of a small group, he later talked about McClendon's "recklessness" and called Chesapeake "one of the worst corporate governance cases" he'd ever seen.

In January 2013, McClendon and Chesapeake announced that McClendon would retire on April 1, 2013—April Fool's Day.

Robert Lawler, the longtime oil and gas executive who took over as Chesapeake's president and CEO in June 2013, later told a crowd at the Houston Producer's Forum luncheon that he felt like Ernest Shackleton, the British explorer who led expeditions to Antarctica in the 1900s. "What I found when I got inside the company was much, much worse than I thought," he said. "It was a really, really challenging dark time." He said that every day he discovered something shocking. "It was beekeepers in gardens, wine collections, and all kinds of crazy things—we are an E&P company," he said.

McClendon, who told a friend that he thought "the board was vulnerable and took it out on him," seemed to have been planning his next act before he walked out the door. Within thirty-six hours of his announced departure, according to a lawsuit Chesapeake later filed against him, he'd asked his assistant

to print out a map of acreage that hadn't yet been leased by
Chesapeake or its competitors in the Utica Shale. On his last
day at the company, he asked another Chesapeake employee to
give him the contact information of the negotiator for one of
the acreage owners. McClendon specifically requested that [the
employee] not provide that information by e-mail," the lawsuit
alleged. The suit would also accuse McClendon of misappropriating information that belonged to Chesapeake on other plays.

McClendon immediately leased office space at the top
of a tall building next door to Chesapeake, where, according
to someone who knows him, he could look down at Chesapeake from his perch. He hired former Chesapeake staff, and
put up a billboard near the entrance to Chesapeake's campus to
announce that his new company was hiring.

Within months, McClendon started a smorgasbord of new
companies under the umbrella American Energy Partners. John
Raymond, who after leaving Howard Weil had gone on to his own
illustrious career in oil and gas, and now ran a highly respected
private equity firm called The Energy & Minerals Group, agreed
to put money into McClendon's new companies, partly because
of the operational team McClendon was bringing with him. The
funding was done only after they'd agreed to a specific business plan, and it was structured so that McClendon couldn't
do much without EMG's approval. Each company was concen-
trated in a specific play, because McClendon planned to take the
companies public within a few years. "The nature of the business," according to a filing much later in probate court, was to
grow the companies "so that they could be spun off and sold or
otherwise monetized." Within just a few years, McClendon had

52   raised $15 billion in capital and his companies employed eight hundred people.

As McClendon was embarking on his second act, the price of gas began to rise, eventually hitting over $6. Experts predicted that oil prices would stay around $100 a barrel. However, it was about to become clear that predicting the direction of energy prices is a fool's errand—and that Aubrey McClendon's destiny was not entirely his to control.

Skeptics

It probably isn't surprising that McClendon, dating all the way back to those early adventures in the Austin Chalk, drew the attention of prominent short-sellers, eventually including Jim Chanos of Kynikos Associates, who is best known for his bet against Enron.

What might be surprising, given all the hype around shale, is that the skepticism extended to the entire industry. "Aubrey McClendon did everything the other guys are doing, just on steroids," says Chanos. "The industry has a very bad history of money going into it and never coming out."

It wasn't until later that the industrywide skepticism burst into the open.

No one would ever mistake David Einhorn for Daniel Plainview, the silver miner-turned-oilman played by Daniel Day Lewis in *There Will Be Blood*, the movie inspired by Upton Sinclair's novel *Oil!* Tall and slightly framed, the baby-faced Einhorn spoke with a high, nasal-inflected voice from behind

54   a podium at the 2015 Ira W. Sohn Investment Research Conference, known as the Super Bowl of the hedge fund industry.
At the Sohn conference in May 2008, he'd made a now-famous
proclamation that the investment bank Lehman Brothers was
in far worse shape than it was letting on. When Lehman went
under, Einhorn's firm, Greenlight Capital, made a fortune from
its giant short position in the company.

People pay thousands of dollars for a seat at the Sohn conference. They come to hear speakers like Einhorn pitch investment ideas, whose impact is often felt on the stock market the
very next morning. Wearing a dark suit and tie, he bounded to
the podium and proceeded to take aim at his latest target: the
shale industry. He proceeded to lambast fracking companies as
financial hucksters and made a detailed case for taking a short
position in a company called Pioneer Natural Resources, which
he dubbed "Mother Fracker."

At the time, Pioneer was run by Scott Sheffield, a former
petroleum engineer who hailed from a family of wildcatters. In
the late 1970s, Sheffield's father-in-law, Joe Parsley, hired him to
work at his company, an independent producer called Parker &
Parsley, which later merged with T. Boone Pickens's Mesa Petroleum to form Pioneer. For years, Pioneer languished, but by the
late 2000s, the company was also drilling in the Eagle Ford shale
formation in South Texas and saw its production surge.

Einhorn's firm had looked at the financial statements of the
sixteen largest publicly traded frackers, which included companies like Pioneer and EOG. Einhorn found that from 2006 to
2014, the fracking firms had spent $80 billion more than they
had received from selling oil and gas. Even when oil was at $100

a barrel, none of them generated excess cash flow—in fact, in 2014, when oil was at $100 for part of the year, the group burned through $20 billion.

A key reason for the terrible financial results is that fracked oil wells in particular show an incredibly steep decline rate. According to an analysis by the Kansas City Federal Reserve, the average well in the Bakken declines 69 percent in its first year and more than 85 percent in its first three years, while a conventional well might decline by 10 percent a year. One energy analyst calculated that to maintain production of 1 million barrels per day, shale requires up to 2,500 wells, while production in Iraq can do it with fewer than 100. For a fracking operation to show growth requires huge investment each year to offset the decline from the previous years' wells. To Einhorn, this was clearly a vicious circle.

Another skeptical investor named Jonathan Tepper, who founded a firm called Variant Perception that provides research to hedge funds and family offices, put together a presentation in which likened the dynamics of fracking to the Red Queen's race in *Alice in Wonderland*: "The Red Queen has to run faster and faster in order to keep still where she is."

Because the Red Queen's race requires so much money, it wouldn't have been possible without the ultra-low interest rate policy that the Federal Reserve has had for the last decade. Amir Azar, a fellow at Columbia's Center on Global Energy Policy, wrote that by 2014, the industry's net debt exceeded $175 billion, a 250 percent increase from its 2005 level. But interest expense increased at less than half the rate debt did, because interest rates kept falling.

56   Einhorn also pointed out that presentations by frackers like Pioneer can be quite misleading. A typical presentation, like the ones by Pioneer, would claim that their wells generate internal rates of return of 40 to 100 percent, which are spectacular numbers. And yet, Pioneer reported negative earnings every quarter through 2015, as did many other companies. Turns out, the financial results from an individual well don't include corporate expenses, such as the money that is spent acquiring land or leases. They also exclude the ongoing capital that needs to be spent in order to maintain production. And that capital literally disappears into a hole in the ground. "Once you extract the oil from the ground," Einhorn said, "That's it. Poof! It's gone."

There's also the Austin Chalk issue. One well might produce stupendous results. But that doesn't mean a well ten miles away will do the same.

Nor are the estimates of how much oil and gas there is in the ground all that meaningful. Companies are required to report their reserves to the Securities and Exchange Commission, but the numbers they report are based on an SEC formula that tries to ascertain how much of the oil and gas in the ground would be profitable to drill. One investor analyzed seventy-three shale drillers in 2014, and found that almost all of them reported higher oil and gas prospects to investors than they did to the SEC. For instance, Chesapeake reported 2.7 billion in "barrels of oil equivalent"—a measure that equates natural gas with oil—to the SEC, but 13.4 billion to investors. Pioneer reported 845 million to the SEC and 11 billion to investors. In total, the

industry reported 33 billion of barrels of oil equivalent to the
SEC and 163.5 billion to investors.

Einhorn, though, was less skeptical of natural gas than of oil.
That's in large part because it takes far less effort and expense
to get natural gas to flow through fracked rock than it does to
get oil to do so, and the decline rates are far less severe. In a 2012
research report, Credit Suisse noted that the average recovery of
a gas well is three to five times that of a typical oil well.

Einhorn's views were mostly echoed by the work done by
a little-known employee-owned firm called SailingStone Capital Partners. Based in San Francisco, SailingStone special-
izes in natural resources investments for its clients—mainly
endowments and family offices—and does exhaustive research,
including hiring engineers to evaluate various plays and digging
deeply into the numbers.

SailingStone had already quietly been raising concerns about
the way executives were paid in the shale business. Instead of
being paid based on the financial results they produced, they were
paid for growing production, regardless of the profits.

After Einhorn's presentation, SailingStone wrote its own
letter to investors. The firm decried the industry's "myopic
obsession with production growth," and concurred with Einhorn that "shale gas looks like a better business than shale oil"
based on historical returns. But the firm did take issue with
Einhorn's blanket dismissals, arguing that not all oil frackers
were alike—for instance, EOG, which Einhorn dismissed as
"Father-Fracker," had reported very different results than Pioneer and others over the full cycle. "The industry is not nearly

58   as uniformly value destructive as David Einhorn suggests," they wrote.

But overall, SailingStone was in agreement with Einhorn. Shale oil and gas firms needed to become better stewards of investors' capital.

Bust

Ali Al-Naimi started his first job when he was four years old, according to his biography *Out of the Desert: My Journey from Nomadic Bedouin to the Heart of Global Oil.* He was a shepherd tending to a flock of lambs for his mother's nomadic Bedouin tribe in the Arabian desert. Unbeknownst to him and nearly everyone else at the time, the ground beneath him housed vast quantities of oil that U.S. companies, in the 1930s, were only just beginning to discover.

Nearly three-quarters of a century later, Al-Naimi had a different job. He was Saudi Arabia's oil minister. Al-Naimi had joined Aramco, the gargantuan state-owned oil company—which according to press reports oversees the production of one in every eight barrels of oil sold worldwide—as a twelve-year-old and rose to become its president. Former U.S. Federal Reserve Chairman Alan Greenspan called Al-Naimi the most powerful man you never heard of. When Al-Naimi spoke, energy markets listened.

60   In late November, 2014, Al-Naimi and his fellow OPEC oil ministers gathered for a meeting in Vienna. The price of a barrel of oil had already begun to slide to below the level experts had expected, to around $80 a barrel. OPEC faced a decision. Its market share was falling, just like in the early 1980s. If the U.S. shale boom continued, OPEC's share would likely shrink further. And so, in the run-up to the meeting, there was an argument that OPEC should once again cut production in order to prop up prices.

But there was a big downside to doing so. While high oil prices benefit producers in the short run, there's an argument that they also speed up the transition to renewables. As one long-time analyst puts it, "With oil there is a price that kills supply and a price that kills demand." And even if OPEC managed to come to an agreement, Saudi Arabia might have difficulty enforcing production cuts among other members. Al-Naimi had once called Saudi's move to cut production in the 1980s, which ended up benefiting others, "an unfortunate decision." This time would be no different. Saudi Arabia's attitude, according to one person familiar with the debates, was, "We're not doing this on our own." Reuters reported that Al-Naimi made special trips to Venezuela and Mexico, and even held a meeting with Alexander Novak, then the head of Russia's state-owned gas and oil giant, Rosneft, probably to negotiate a joint production cut. But a production cut would play into the hands of U.S. frackers, who would almost certainly gain market share at OPEC's and Russia's expense.

On the other hand, if prices continued to fall, U.S. frackers would suffer far worse, the OPEC ministers reasoned. It costs a lot more to produce a fracked barrel of oil in the U.S. than it does to get a barrel of oil out of the ground in Saudi Arabia and other OPEC countries. In a rare interview with the *Middle East Economic Survey*, Al-Naimi said that Saudi production costs are no more than $5 per barrel, and that marginal costs of development are "at most" $10 per barrel. As 2014 drew to a close, estimates were that it cost U.S. frackers as much as five times that to get a barrel out. "Is it reasonable for a highly efficient producer to reduce output, while the producer of poor efficiency continues to produce?" asked Al-Naimi. "That is crooked logic. If I reduce, what happens to my market share? The price will go up and the Russians, the Brazilians, U.S. shale oil producers will take my share." Al-Naimi and OPEC thus came to the decision to leave production levels where they were.

Those who know Saudi Arabia caution that it's impossible for outsiders to know what the real reason for anything is, and that it's quite possible there was more than one motive. Some experts argue that Saudi's painful history with production cuts is a better explanation than any conspiracy theory. But the Thanksgiving Day decision was interpreted as an attempt to put American frackers out of business. "Inside OPEC Room, Naimi Declares Price War on U.S. Shale Oil," announced a Reuters headline the day after the meeting. "Just because you're paranoid doesn't mean they are not out to get you," said economist Erik Norland. "This was definitely an attempt to kill the U.S. frackers."

But, as the longest period of high oil prices in history came to an end, it's safe to say that absolutely no one expected what would happen next. After OPEC's Thanksgiving Day decision, oil prices skidded more than $6 a barrel. By the end of the year, the price was less than $60, and by February 2016, a barrel of oil fetched just $26.

As Al-Naimi said, it's not difficult to see that propping up oil prices would benefit U.S. oil, and that the opposite would quickly expose the weak underbelly of U.S. shale—its high costs and ravenous need for capital.

The reckoning arrived soon. Once-booming U.S. production hit the skids. The so-called rig count—the number of rigs drilling for oil and gas at a given time—fell from 1,920 rigs in late 2014 to a low of 480 in early 2016. "We think it likely that to find a lower level of activity would require going back to the 1860s, the early part of the Pennsylvania oil boom," Paul Hornsell, head of commodities research for Standard Chartered Bank, wrote in a research note. By mid 2016, U.S. oil production had declined by a million barrels a day.

One after another, debt-laden companies began to declare bankruptcy, with some two hundred of them eventually going bust. Samson Resources declared bankruptcy in the fall of 2015; in a legal filing, the company estimated its value at less than $2 billion, a fraction of the $7 billion KKR had paid. The bankruptcy totally wiped out the $4.1 billion in cash KKR and its partners had invested in the company.

Halcon, the company Floyd Wilson had started after selling Petrohawk, went belly-up in the summer of 2016. A company called Quicksilver Resources also filed for bankruptcy, listing $2.1 billion in debt. Its assets later sold for just $245 million.

In a report released in the fall of 2016, credit rating agency Moody's called the corporate casualties "catastrophic." "When all the data is in, including 2016 bankruptcies, it may very well turn out that this oil and gas industry crisis has created a seg-mentwide bust of historic proportions," said David Keisman, a Moody's senior vice president.

Some of those who had bought assets from McClendon and others in the heyday also began to write down the value of what they'd purchased. Statoil, the Norwegian energy giant, wrote down the value of its shale and Canadian oil sands assets by $4 billion; Royal Dutch Shell reported a write down of more than $8 billion. Most prominent was Australia's BHP Billiton, which had spent $5 billion investing with Chesapeake in the Fayette-ville shale and plowed another $15 billion into the purchase of Houston-based Petrohawk. BHP put all the assets on the block in the fall of 2014, but found no buyers, and eventually wrote off over $7 billion—which begat the phrase "pulling a BHP."

As one investor put it: "All of the acquisitions of shale assets done by the majors and by international companies have been disasters. The wildcatters made a lot of money, but the companies haven't."

As shale companies slashed their budgets, fracking equip-ment was idled—according to research firm IHS Markit, close to 60 percent of the fracking equipment in the U.S. was inac-tive. Shale companies and oilfield service companies laid off workers. All told, the global oil and gas industry shed almost half a million jobs during the bust, according to consulting firm Graves & Co.

The shale boom towns suddenly resembled their Cali-fornia counterparts after the Gold Rush. In the Cline shale east of Midland, Devon decreased its rig activity and let its leases

64 expire, citing "a lot of variability" in the formation; its partner, Japan's Sumitomo Corp, took a $1.55 billion write-down on its Cline investment. The town's "ambitions are fading fast as the plummeting price of oil causes investors to pull back, cutting off the projects that were supposed to pay for a bright new future," wrote The Associated Press in early 2015. "Now the town of 11,000 awaits layoffs and budget cuts and defers its dreams."

By nearly all accounts, the shale boom had gone bust. In early 2016, non-investment grade energy bonds—the shale industry's rocket fuel—yielded 25 percent, five times what they had a year and a half earlier. "This has the makings of a gigantic funding crisis" for energy companies, William Snyder, the head of Deloitte's U.S. restructuring unit, told the *Wall Street Journal* in early 2016. That spring, the Kansas City Federal Reserve concluded that "current prices are too low for much long-term economic viability of shale oil production."

Around the same time, Pioneer's CEO Scott Sheffield told CNBC, "I see half the independents going into Chapter 11 or bankrupt if this thing lasts another one to two years . . . [the price of a barrel of oil] really needs to get back up to $50 to $60 to have these companies survive."

Surveying the carnage in the spring of 2016, then-ExxonMobil CEO Rex Tillerson told a gathering of analysts that due to the huge amount of debt most companies in the industry had accumulated, he couldn't even find anything worth buying.

It Changes the World, but It Ends in Tears

As shale's top evangelist, Aubrey McClendon was having a time of it. One of the many new companies he'd founded, called American Energy Partners, was looking to raise $2 billion. It wound up gathering only $11.3 million before the IPO was canceled. McClendon tried to start another "blank check" company called Avondale Acquisition Corp, which also sought to make oil and gas acquisitions. That too failed to get off the ground. McClendon's world seemed in tatters. By early 2016, bonds sold by some American Energy entities had plunged to about fifteen cents on the dollar, according to Trace, a bond-price reporting system of the Financial Industry Regulatory Authority.

Even in those dark days, McClendon remained a true believer. Rather than back down he doubled down. He announced deal after deal: the purchase of 55 million acres of oil and gas properties in Australia; a joint venture with YPF, Argentina's national oil company, to explore the Vaca Muerta,

66   or Dead Cow, shale field; a partnership with Mexican companies to explore that country's resources; and more. Even those who were skeptical of him were amazed. "Look at this guy who mortgaged it all to start a new company in the teeth of a terrible decline," says one financier. "If he went out, he was going to go out in a blaze of glory."

It was no exaggeration to say McClendon "mortgaged it all." Legal filings offer a glimpse into some of the frantic fundraising propping up his tottering empire. In the fall of 2014, he took out a $465 million loan from a handful of Wall Street banks, including Goldman, as part of the fundraising for his American Energy Partners companies. He also secured that loan with a personal guarantee, meaning that everything from his homes to his wine collection to his antique map collection to his part-ownership of the Oklahoma City Thunder effectively was hocked, maybe more than once. The *Wall Street Journal* later reported that McClendon was listed as a debtor in nearly a hundred financing statements and collateral agreements.

In the fall of 2015, he borrowed $85 million from Oaktree, a distressed debt investor, in order to buy 18,000 acres of oil and gas leasehold interests in a part of South Central Oklahoma called the Stack/Scoop. Court documents later showed that McClendon personally got $19.7 million of the loan proceeds "as an advance" of his share of eventual profits. That loan, too, was secured by his stake in the Oklahoma City Thunder—a stake that had previously been pledged to Bank of America but reassigned to Oaktree. "When he had assets, he put leverage against them," one banker says. (After his death, the Stack/Scoop interests were sold, and Oaktree was repaid.

Given McClendon's personal guarantee to the Goldman group of lenders, as well as personal guarantees on the EIG loans, it wasn't clear where the money from the sale of the Thunder, which some creditors regarded as his most valuable asset, would end up.)

Even his wife got into the act. Oilmen in Midland say that when McClendon was trying to close a deal, his wife chartered a plane, flew out, and signed a check, letting everyone know that the $10 million was hers—the McClendon family was all in.

But the economics weren't working anywhere. McClendon also lost control of the companies he had set up to manage his interests in Chesapeake's wells. One knowledgeable source says that the returns at the wellhead, meaning before expenses such as corporate overhead and the cost of transporting the gas, were only in the high-single-digits. The cash McClendon was getting out of his share of the wells didn't cover the cost of his loans and the additional funds he had to pony up to cover his share of the drilling costs. The deals were restructured multiple times, but eventually, his main lender, EIG Global Energy Partners, took over between 70 percent and 85 percent of McClendon's companies, according to later court filings.

Those who know McClendon and who backed him believe he might have survived the financial hell, maybe even raised the capital for a third go round. But he could not escape the legal hell he also found himself in.

McClendon's legal nightmare predated the shale bust. In the spring of 2014, a year after McClendon had left Chesapeake, the state of Michigan brought criminal charges against the firm for conspiring with other companies to rig the bids in a 2010

68   state auction for oil and gas rights. Michigan soon added felony racketeering and fraud charges, accusing Chesapeake of systematically swindling individual landowners. The alleged conspiracy dated back to May 2010, when the state of Michigan had auctioned off a chunk of state-owned land to oil and gas drillers; McClendon and the CEO of a Canadian company named Encana had divvied up the state, agreeing not to bid on leases in each other's allocated counties, according to the charges. "Should we throw in 50/50 together here rather than trying to bash each other's brains out on lease buying?" McClendon once asked an Encana executive, according to evidence filed in the case.

A raft of civil lawsuits were filed alleging that Chesapeake had used a similar strategy in other hot plays. In these complaints, McClendon's co-conspirator was alleged to be his old partner, Tom Ward, who once told a trade publication that he and McClendon had partnered all those years ago after realizing that "we would be better off sharing," and who after leaving Chesapeake had started a new company called Sandridge. At weekly meetings, the companies agreed on offers that their landmen would make in the upcoming week, one suit alleged. "Senior officials of both companies, including McClendon and Ward, knew about and condoned these meetings and the bidding practices that resulted from them."

The bid-rigging allegations were a huge irony given McClendon's reputation for overpaying.

In April 2015, Chesapeake paid $25 million to settle the Michigan charges. It looked as if Chesapeake, which told shareholders that it had done its own investigation and found no wrongdoing, had made the matter go away.

But what hadn't gone away was the animosity between Chesapeake and McClendon. In early 2015, Chesapeake sued McClendon, alleging that he'd taken confidential information when he left, and used that information to set up American Energy Partners. (At the time, McClendon said in a statement that it was "beyond belief" that Chesapeake had "decided to add insult to injury almost two years to the day after his resignation by wrongly accusing him of misappropriating information.")

In fact, Chesapeake had thrown McClendon under the bus. When Chesapeake settled the charges, its lawyers used something called the Conditional Leniency Program, which shielded the company from criminal antitrust charges, fines, and penalties—but in order to be admitted to the program, Chesapeake had to admit that there had been a criminal violation of the antitrust laws, which the company blamed on its former CEO.

McClendon knew he was under criminal investigation. American Energy Partners even disclosed in a financial filing that the Justice Department had launched a formal inquiry. What made it even more of a mess is that McClendon had personally guaranteed many of his debts, and according to someone familiar with events, those guarantees allowed him to be sued if he was accused of criminal activity.

But if McClendon was fazed, he didn't show it. In the fall of 2015, McClendon hired famed trial attorney Abbe Lowell. One investor, who took him to a basketball game in the spring of 2016, says, "Even with his world crumbling around him, he was always a promoter and ever the optimist."

70   By that time, business publications were reporting that John Raymond was also pulling his support from McClendon. Initially, EMG had defended McClendon from at least some of the allegations. For instance, EMG called Chesapeake's lawsuit meritless. But then, a few months after the suit was filed, one of the American Energy Partners companies announced that it had agreed to assign approximately 6,000 acres in Ohio to Chesapeake and pay it up to $25 million; Chesapeake in turn agreed to drop the company and some 20 investors from the lawsuit.

Then the hammer fell. At 5:30 p.m. on March 1, a federal grand jury in Oklahoma City indicted McClendon for a "conspiracy to rig bids" that existed from late 2007 through at least March of 2012. Although the other party to the conspiracy wasn't identified or indicted, it was Tom Ward. McClendon's "actions put company profits ahead of the interests of leaseholders entitled to competitive bids for oil and gas rights on their land," said William J. Baer, assistant attorney general in the Antitrust Division.

That night, according to Reuters, McClendon had been expected at a private dinner with potential business partners including Vicente Fox, the former president of Mexico. McClendon never showed, though the dinner guests opened three bottles of wine from his legendary collection, including a 2010 Napa Valley red bearing American Energy Partners' logo.

When the indictment became public the next morning, McClendon had a statement ready. "The charge that has been filed against me today is wrong and unprecedented," he said. "Anyone who knows me, my business record and the industry in which I have worked for thirty-five years, knows that I could not be guilty of violating any antitrust laws.... I am proud of my track record in this industry, and I will fight to prove my innocence and to clear my name." At around the same time, Raymond's firm, reported Reuters, sent its investors a letter informing them that the firm would "cease any and all new business activities" with McClendon, who as of that February was no longer the CEO or a board member of any of its portfolio companies. Raymond wrote that it "purely a coincidence" that the charges were brought at the same time as "business arrangements" were being finalized, but he also noted that "these are serious allegations that have been made against McClendon (and could have equally serious implications across the industry.)"

Around 9:00 a.m., McClendon, known for driving fast and without a seat belt and for multitasking on his phone, left the office to meet someone for breakfast at Pops, which he owned, having transformed it from a gas station on the historic Route 66 into an ultra-modern take on the roadside restaurant, with the slogan "Food, Fuel, Fizz" and a 100-foot cantilever roof that's won multiple architecture awards. According to later police reports, he was traveling in his Chevy Tahoe at almost ninety miles per hour—well over the posted speed limit of fifty miles an hour—when his car collided with a concrete wall supporting a highway overpass at 9:12 a.m. The Tahoe burst into flames.

A series of 911 calls described the scene:

"It looks like a Tahoe and it looks pretty rough . . ."

"The cab is completely crushed . . ."

"That vehicle just exploded . . ."

The front end of McClendon's vehicle had hit the overpass support head-on, leading to initial speculation that it was

72 a suicide. McClendon drove "through a grassy area right before colliding into the embankment," Captain Paco Balderrama of the Oklahoma City Police told reporters. "There was plenty of opportunity for him to correct and get back on the roadway, and that didn't occur."

But after an investigation that included interviews with McClendon's friends and associates, the police did not find anything that suggested he was seeking to end his own life. On June 8, the state medical examiner ruled that McClendon's cause of death was an accident. He was fifty-six.

It was hard not to see McClendon's death as the punctuation marking the end of an era. As Australian hedge fund manager John Hempton asked, "Is Chesapeake the model for this business? It changes the world, but it ends in tears?"

Saudi America

Part Two

America First

In a great irony, without the oil bust, it's unlikely that the most symbolic development of all—the lifting of the export ban—would have happened. And without ultra-low natural gas prices in the U.S., our wealth of natural gas wouldn't be the global game changer that it may become, either.

American policymakers have always recognized the power of energy not just to reshape the U.S. economy but also to remake geopolitics. Recall the handmade "Temporarily Closed . . . OUT OF GAS" signs at deserted filling stations across the U.S., among the most iconic images from the 1970s, speaking powerfully to a nation's newfound sense of vulnerability and helplessness.

Fracking promised the beginning of a new era of American energy abundance not seen since the formation of OPEC. In the fall of 2009, the IEA's chief economist, Fatih Birol, told the Council on Foreign Relations that, "There is a silent revolution taking place in the United States, so silent that nobody's aware of it." Citigroup chief economist Ed Morse said that the U.S. had the potential to become the "new Middle East," and

Leonardo Maugeri, a former director at Italian energy firm Eni   75
who became a fellow at the Harvard Kennedy School's Belfer
Center for Science and International Affairs, coined the phrase
"Saudi America" when his 2012 report predicted that the U.S.
could one day rival Saudi Arabia's fabled oil production.

For many years, oil executives hadn't even contemplated
exports, for the simple reason that there wasn't anything to
export. But as that changed, there was a burgeoning push, led
by a coalition of energy industry CEOs including Harold Hamm
of Continental Resources, and Ryan Lance, the CEO of ConocoPhillips, to allow oil exports. They found allies in Washington, although for a long time, those allies were very quiet.

The first public salvo came in the summer of 2013, when the
Council on Foreign Relations published "The Case for Allowing
Crude Exports." CFR pointed out that for the first time in over
sixty years, the U.S. had become a significant gross exporter
of refined oil products, like gasoline and diesel. Refined products weren't subject to the ban, so why not lift the entire ban?
CFR's main argument was that doing so—this being before the
age of Trump and its attendant backlash against globalism—
would demonstrate America's commitment to free trade. And
CFR argued that while "proponents of the ban might argue that
it [the ban] increases national security by slowing the depletion
of U.S. oil fields," removing the ban would actually increase our
security because it would catalyze production.

At the same time, there was a growing awareness that this
was about more than oil. In 2012, President Obama sounded
almost like Aubrey McClendon in his annual State of the Union
speech: "We have a supply of natural gas that can last America

76 nearly a hundred years, and my administration will take every possible action to safely develop this energy," he said.

Although natural gas doesn't occupy the same place in the national psyche that oil does, it is a potent geopolitical force. Back in 1981, advisers in the Reagan Administration warned that if the proposed Trans-Siberian pipeline were to be built, crossing modern-day Ukraine and bringing natural gas from Russia to Europe, it would foster European dependence on Russian fossil fuels. But it was built, and by now, according to analysis by J.P. Morgan, six Baltic and Eastern European countries rely entirely on Russia for their gas supplies; Germany gets 40 percent of its gas from Russia. As tensions between Russia and Ukraine began to flare starting in the mid-2000s, Europe's complacency changed to panic.

During the Obama Administration, U.S. and European politicians began pushing for America to accelerate the granting of permits for new LNG facilities so that the U.S. could export natural gas to Europe, weakening Russia's ability to use its energy supplies as a political weapon. The ability to "turn the tables and put the Russian leadership in check lies right beneath our feet in the form of vast supplies of natural energy," John Boehner wrote in a March 2014 *Wall Street Journal* op ed. Ambassadors from Hungary, Poland, Slovakia, and the Czech Republic sent a letter asking Congress to allow the faster sale of more natural gas to Europe.

"There was a deep thirst and interest and deep concern around the topic of LNG," recalls a former Obama Administration official. "Country after country asked, 'When are the brakes coming off?'" But, he says, "there were a lot of concerns about the

unintended consequences if you used energy politically. . . . It's
hard to tell Russia to knock it off if we're doing the same thing."

Even the *New York Times* weighed in on the side of exports.
"The benefits of selling gas to other countries would more
than offset the modestly negative impact of higher prices for
domestic users of the fuel," opined the paper's editorial page.

There were indeed fears that exports would cause prices
to rise—but the bust made those fears less potent. In 2012, the
U.S. approved the first construction of an LNG plant, to be built
by Cheniere Energy at Sabine Pass, on the border of Texas and
Louisiana.

Exporting natural gas requires the Energy Department to review
LNG export permit applications in order to ensure that they are
in the nation's best interest, but exporting oil required Congressional action. The first member of Congress to call for a repeal
of the ban was Republican Senator Lisa Murkowski of Alaska,
who did so in early 2014. She was soon joined by a chorus of
voices, such as *New York Times* columnist Thomas Friedman,
who argued in an op-ed that "nothing would make us stronger
and Putin and ISIS weaker." His point was that lifting the ban
would "significantly dent the global high price of oil," thereby
weakening regimes that depended on high oil prices.

Environmentalists were adamantly opposed to the repeal,
with the Sierra Club arguing that it would increase oil drilling and "create yet another consumer giveaway to an already
wealthy industry" by causing the price of gasoline to rise. Even
the powerful American Petroleum Institute, long the chief lobbying organization for oil and gas producers, was split, because

78   some refiners opposed to lifting the ban—the lower the price they had to pay for crude, the higher their profit margin.

In the spring of 2014, the refiners even formed their own lobbying group called Consumers and Refiners United for Domestic Energy, or CRUDE. "We're on the cusp of a historic opportunity—finally—to gain energy independence and security, and break through the grip of foreign oil cartels on the U.S. economy," the lobbyist for CRUDE said. "To smash that opportunity away by all of a sudden exporting crude oil is definitely not in the interest of the United States."

Later that year, a group of producers countered with their own lobbying group, called Producers for American Crude Oil Exports, or PACE. The coalition hired a longtime lobbyist named George Baker of Williams and Jensen with a single goal in mind: to overturn the ban. Among other things, industry-friendly groups produced studies showing that exports would decrease, rather than increase, the price of gasoline, because exports would help lower the global benchmark price.

In response to the push, the Obama Administration was publicly cagey, saying only that it was a "policy decision" that should be made by the Commerce Department, so the White House wouldn't support legislation specifically aimed at repealing the ban. When the House passed a bill, President Obama threatened to veto it, arguing that it would further American reliance on fossil fuels.

Yet those pushing for exports started to get their way in the summer of 2014. The ban was not revoked, but a certain kind of product called "condensates," which in essence are any types of

oil that condense from gas into a liquid after being set free from   79
a high pressure well—were reclassified by the Obama Administration as a refined oil product, and refined products weren't
subject to the ban. "'When does it stop being crude oil?' was the
new parlor game," says a former Obama Administration official.

The truth was that many in the administration were in
favor of lifting the ban. The issue was appeasing constituents,
especially environmentalists, who opposed any policy change
that might increase domestic oil production. "There was not
a man jack in that administration who didn't understand the
argument, whether it was people at State, at Defense, or at the
National Economic Council," says Baker. "Sotto voce, they'd say,
'We get it.'" So, he says, instead of using Republicans to build
an argument, PACE turned to former Obama Cabinet members,
from Leon Panetta, who had served as the Secretary of Defense
and the Director of the CIA under Obama, to Larry Summers,
who had served as the director of the NEA. "We had a whole
Congressional hearing that was all supportive testimony from
former Obama Administration officials," Baker says.

Summers, for his part, told an audience at the Brookings
Institution, "The merits are as clear as the merits with respect to
any significant public policy issue that I have ever encountered."
Panetta co-authored a 2015 *Wall Street Journal* op-ed with Stephen Hadley, who had served as the National Security Advisor
under President George W. Bush, in which they argued: "The
U.S. remains the great arsenal of democracy [and] it should also
be the great arsenal of energy." Democrats from oil- and gas-rich
states, like Senator Heidi Heitkamp from North Dakota, were
also in favor of lifting the ban.

As prices began their plunge in the fall of 2014, and producers had to lay off workers, the pressure to remove the ban
only intensified. Studies came out from various industry-
friendly organizations arguing that free trade in oil would create
nearly a million jobs, add billions to the economy, and would
lower the trade deficit. "One of our messages was to present the
issue as it properly is, not just about oil and gas but about the
much broader social and economic benefits to the nation," says
Baker.

At the same time, environmentalists were quieter than they
might otherwise have been, because amid the bust, exports no
longer seemed like such a big deal.

Baker, who has been lobbying since 1980, says he knew all
along that repeal would never be passed on a stand-alone basis,
but rather as part of an omnibus bill. Repeal of the export ban
was ultimately tucked into the sprawling $1.1 trillion year-end
2015 spending bill. In exchange, Democratic lawmakers got
extensions on tax credits for wind and solar power that were
due to expire. "In this age of things not getting done, this was a
throwback to the era that almost no longer exists, to building a
community of interests in the spirit of compromise," Baker says.

But if you weren't somehow invested in the ban or its repeal,
then you probably didn't even realize what had happened. The
legislation passed the House, and then the Senate, before noon
on December 18, 2015. By the early afternoon, President Obama
had signed it. Then everyone raced out of Washington for the
holidays. "By 6:00 p.m., no one was left in town," Baker says. "To
celebrate, I went to Lia's in Bethesda and had a Manhattan and a
steak by myself. I thought it was a big, big, big deal."

Permania

While on the campaign trail, then-candidate Donald Trump began to talk about energy independence. Upon election, he installed one of the most energy heavy cabinets in modern history, from ExxonMobil CEO Rex Tillerson as Secretary of State; to former Oklahoma Attorney General Scott Pruitt as head of the Environmental Protection Agency, which he had sued more than a dozen times to protect the interests of energy companies; to former oil and gas consultant Ryan Zinke as Secretary of the Interior. (Before his death, Aubrey McClendon had made a $10,000 donation to a PAC formed by Pruitt, who was then Oklahoma AG, to raise money for other candidates, according to Oklahoma Watch, which is sometimes the mark of someone with ambitions for higher office.)

After the election, President Trump upped the ante. In May, he promised "complete" independence from foreign sources of oil, saying, "Imagine a world in which our foes and the oil cartels can no longer use energy as a weapon. Wouldn't that be nice?" In

82   June, he took it a step further, saying to a group gathered at the Department of Energy for an event called "Unleashing American Energy," "We are really in the driving seat. And you know what? We don't want to let other countries take away our sovereignty and tell us what to do and how to do it. That's not going to happen. With these incredible resources, my administration will see not only American energy independence that we've been looking for so long, but American energy dominance."

By the time Trump took office in January 2017, oil was booming again, and the place to see that was Texas.

Thirty-five floors above Houston, the city's new Petroleum Club was filled with sun on late spring afternoon in 2017. About two hundred people from the Texas energy elite were gathered in a space infused with what its interior designer described as "*Mad Men* style." They were there to celebrate Cody Campbell and John Sellers, who had just sold their company, Double Eagle Energy, to another fracker called Parsley Energy for $2.8 billion.

Campbell had played for the Indianapolis Colts until an injury ended his career. He and Sellers were high school buddies from Canyon, Texas who worked in real estate until the 2008 financial crisis. At that point, they switched to oil and gas, and used bank loans and money from friends and family to start buying up leases directly from landowners—much as Aubrey McClendon once did—in an area of west Texas called the Midland Basin. Part of a larger, longtime oil region known as the Permian Basin, the Midland Basin encompasses 75,000 acres that stretch across the southeast corner of New Mexico and western Texas. The urban center, such as it is, of that

region has always been the city of Midland. As *Texas Monthly*
once wrote. "People in Midland like to say that God felt such
remorse about what he did to the land out there that he decided
to give it oil."

Double Eagle Energy wasn't glamorous. "We started out,
just the two of us, just a couple of guys running title and picking
up leases, signing deals on the hoods of trucks," Sellers tells the
room. But back in 2012, with the price of oil around $100, a New
York-based private equity firm, Apollo Global Management,
heard about what they were doing and offered to back them.
Just before the crash, in 2014, Sellers and Campbell sold some
of the leases they'd accumulated in an area of Oklahoma called
the Scoop & Stack to McClendon's American Energy Partners
for $251 million.

They kept going, and took advantage of the bust to scoop
up more leases at discounted prices. Now in their mid-thirties,
they've already made fortunes, but they still have the down-
home easiness of West Texas. As Sellers is speaking, another
guy in the room whispers to me, "You have to like a guy who just
made some $200 million who is like this."

Campbell and Sellers promptly formed a new company,
again with Apollo's backing, to pursue more Permian Basin
investments. "Cody and I are thirty-five," Sellers says. "We
aren't going to hang it up just yet."

As it turns out, the demise of fracking that seemed so inevitable
wasn't inevitable after all. There are a few reasons why that was
the case, but it all begins with the Permian—Permania, some are
now calling it.

84   Fly into Midland, and all you see across the flat dry land are windmills and drilling rigs. "Outside of Saudi Arabia, the whole oil story today is West Texas," one investor tells me.

It certainly wasn't a secret that there was oil in the Permian. The first oil boom there took place almost a century ago, in the 1920s, when thousands began flocking to Midland. Oil from the Permian fueled the Allied Forces during World War II. In the heyday of American oil that followed, scores of office towers went up in Midland, including the twenty-two-story Wilco building. For many years, it was the tallest building around, and helped give Midland its nickname, "Tall City"—relative to the vast emptiness that surrounds it. In the 1980s boom that followed the oil embargo, newcomers to the city were living in tents, cars, and trailers, according to the Texas State Historical Society; eight midland oil men made the very first Forbes 400 list, which was hugely impressive for a town of 70,000 people. Then the hungry side of the industry's feast or famine cycle set in.

In the decades between then and now, small independent drillers continued to work the tired vertical wells, mainly in an area called the Spraberry Trend, but the majors mostly abandoned West Texas. Drilling a Spraberry well was "like watching paint dry," one oilman told *Texas Monthly* in 2010. "You know where to drill, you drill, you eventually get your ten or so barrels of oil a day, and then you drill another one." In 1999, when the price of crude fell again, a mere 43 rigs were working the area, said the magazine. "Nobody thought the Permian would be the darling of the ball," a Midland billionaire tells me. "The Permian was the girl you could call Friday night and get a date."

By the mid-2000s, the independents were trying horizontal drilling and hydraulic fracking on areas like the Spraberry and the Wolfcamp, a layer of rock that runs right below the Permian and was long thought to be impermeable. It turned out to be a bonanza, the likes of which the world has seldom seen: By 2010, people in Midland were using the "B" word—boom—again. What set the Permian apart from other plays is geological luck. Its oil- and gas-bearing rocks are laid down in horizontal bands. As one engineer explains to me, "Instead of just having one carpet, it's like seven or eight carpets are stacked up." That means that one lease can give you multiple layers of hydrocarbons, and also that you can drill more efficiently, because you only have to use one expensive rig to access multiple layers.

In addition, other celebrated plays like the Bakken and the Marcellus suffered from a lack of infrastructure, most notably pipelines, to process and carry away the oil and gas. It doesn't matter how much you can produce if you can't get it to customers. By contrast, infrastructure from pipelines to facilities that separate oil into its byproducts, were mostly already in place in the Permian.

In 2010, the Permian Basin was producing just shy of 1 million barrels of oil a day. In 2017, that had more than doubled to over 2.5 million barrels a day. By August, output from the Permian alone exceeded that of 8 of the 13 members of OPEC, according to Bloomberg. The International Energy Agency predicts that output will hit more than 4 million barrels a day within a few years. Production from the Permian is the primary driver behind skyrocketing estimates of how much oil the U.S. will produce. In its most recent forecast, the EIA predicted that U.S. crude production would average almost 10.6 million barrels a day in 2018, and 17 million barrels a day by 2023.

In 2016, Saudi Arabia averaged 10.5 million barrels a day, and Russia averaged almost 11 million. America seemed on its way to once again becoming the world's biggest energy power.

The story, believers say, is technology. "Shale 2.0," the industry calls it.

Until recently, the history of shale drilling was that operators would watch what others did, and if someone's new technique got more oil or gas out of the ground, then everyone else would start doing that. "The early science was sometimes no more sophisticated than, 'Look at what Jim over there is doing!'" says Gary Sernovitz, a venture capitalist and the author of *The Green and the Black: the Complete Story of the Shale Revolution, the Fight Over Fracking and the Future of Energy.*

But now, everyone is searching for data-driven, repeatable ways to maximize the amount of oil you can get out of the ground while minimizing the cost. It used to be that a well would travel horizontally for about a mile. By 2015, the Federal Reserve reported that the average was two miles; Chesapeake, Exxon Mobil and Continental have all neared or broken the three-mile mark. In late 2017, Bloomberg reported that one company had drilled a well that stretched almost four miles, longer than the tallest peaks on five of the world's seven continents.

Other measures have gotten grander, too. In 2011, the average well used about four million pounds of sand, says Samir Nangia, the director of energy consulting at IHS Markit. Now, the average is twelve million pounds, and some leading edge

wells use over thirty million pounds. There's a whole new science around the logistics of getting sand from Wisconsin, where
most frac sand is mined, to the job site.

At the same time, other things are getting more minute
and efficient. Drillers are also executing smaller, more complex,
and more frequent fractures. These more precise fracks reduce
the risk that the wells "communicate"—that one leaks into
another, rendering them inoperable—so the wells can be drilled
more closely together. Operators are also utilizing so-called
"pad drilling," where a rig drills multiple wells from the same
spot, and "walking rigs," where a rig can move a few feet without
having to be demobilized. That reduces cost, as does anything
that shrinks the amount of time it takes to drill a well.

According to a 2016 paper by researchers at the Federal
Reserve, not only are rigs drilling more wells, but each well is
producing far more. The extraction from the new wells in their
first month of production has roughly tripled since 2008.
Break-even cost—the estimate of what it costs to get a barrel
of oil out of the ground—has plunged. Before the bust, it was
supposedly around $70; analysts say it's less than $50 now, and
some insist that in certain areas of the Permian, it's as low as
$25, or even $15.

On the surface, Permania doesn't seem to have changed Midland. Unlike in the 1980s, the city doesn't show its hustle or its
wealth. The streets are empty and billionaire oil men are mostly
tucked away in utterly unprepossessing office buildings of the
sort you might find in any small, dying downtown. On the drive to
the golf course, pump jacks look like bleached bones in the mid-
day sun and a billboard reads "God's word to America: Repent."

But in front of the Petroleum Club in Midland, an electronic display flashes the two pieces of information upon which this region's health depends: the oil price and the number of drilling rigs. Right now, there's a steady stream of young geologists, engineers, and executives moving to Midland. Out of the spotlight, landowners who have sold or leased their lands have made fortunes, especially those in the northern Midland Basin, where land is going for over $1 million an acre. One private equity investor tells me that there is a waiting list for private schools and for membership in the Racquet Club, long a center of Midland social life. Fittingly enough, Permian shale runs beneath the grounds of the Racquet Club, which owns the royalties, making it one of the richest country clubs in America.

The dramatic rebound made skeptics look spectacularly wrong—no one more so than David Einhorn. From the time that he recommended shorting Pioneer in the spring of 2015 through early 2017, the company's stock soared some 30 percent to over $175 a share.

As it happened, Pioneer, whose new, modern $50 million headquarters stands out in the dusty brown of Midland, got lucky too. All those years ago, when Scott Sheffield came to work at Parker & Parsley, the company's primary holdings were in the Permian Basin—and he never sold, even through those dreadfully boring decades. When the boom began, Sheffield became the evangelist in chief for the Permian, calling it the "crown jewel of the world's oil and gas industry." In Pioneer's presentations, he told investors that he thought the Permian shales could hold 75 billion barrels of oil, second only to Saudi Arabia's gigantic Ghawar field.

At the end of 2016, Sheffield retired, but the Sheffield name is now its own West Texas dynasty. Sheffield's son Bryan, who had been living in Europe, came home to West Texas to operate some of his grandfather's wells as the boom began. "I was living in Spain, trading commodities, married to a beautiful Spanish girl I had met on the beach, and suddenly it just hit me," Bryan Sheffield told *Texas Monthly*. "I told her we had to move back to Midland because I just had to see what I could do."

The younger Sheffield's timing was impeccable. Although he'd had next to no experience of his own in the oil business, in the spring of 2014, at the height of the boom, he took his company, named Parsley Energy after his grandfather, public. That made Bryan Sheffield one of the youngest billionaires ever in the energy business.

Optimists argue that the Permian is still in early innings—in large part because they believe that technological improvements will continue to slash the costs of drilling a well while boosting the amount of oil that comes out. "Every time people say it can't go down further, companies figure out a way of doing it cheaper," says Nangia. "You are still only extracting about 12 percent of the total hydrocarbons, so there is plenty of room to increase, particularly for oil," he adds. "They will always underestimate," a longtime Midlander says about the forecasts. "And we'll never get the last barrel of oil out of the Permian. It just won't happen."

And yet, even today, it is unclear if we will look back and see fracking as the beginning of a huge and lasting shift—or if we will look back wistfully, realizing that what we thought was transformative was merely a moment in time. Because in its current financial form, the industry is still unsustainable, still

90 haunted by McClendon's twin ghosts of heavy debt and lack of cash flow. "The industry has burned up cash whether the oil price was at $100, as in 2014, or at about $50, as it was during the past three months," one analyst calculated in mid-2017. According to his analysis, the biggest sixty firms in aggregate had used up an average of $9 billion per quarter from mid-2012 to mid-2017.

It wasn't just American technological prowess that helped restart the oil boom.

One unsexy and unheralded factor is so-called service costs—the costs to rent a rig, hire the crew, purchase the sand and other ingredients necessary to frack a well, and so on. Service costs are cyclical, meaning that as the price of oil rises and demand for services increases, the costs rise too. As the price of oil falls and demand dwindles, service companies slash to the bone in an effort to retain what meager business there is.

One investor hired a geologist to perform an analysis of how much the reduction in service costs contributed to the reduction in break-evens; according to this analysis, almost half the reduction was due to the plunge in service costs.

But a reduction in service costs is a temporary phenomenon; as drilling came back, those improvements began to reverse. In the spring of 2017, the CEO of SLB, a big service provider, told a gathering of investors that "there is an impending cost inflation avalanche coming from the service industry." By the spring of 2018, there were rumors that truckers were getting $1,000 signing bonuses.

"What people still fail to understand is that the most cyclical number we have is the theoretical break-even," one longtime oil man says. "There will be stories about how the $40 break-even became the $70 break-even, and people will say, 'Who lied to me?'"

And so it is that the most important factor in the comeback of shale is the same thing that started the boom in the first place: The availability of capital. "It came back because Wall Street was there," says longtime short-seller Jim Chanos. In 2017, U.S. frackers raised $60 billion in debt, up almost 30 percent since 2016, according to Dealogic.

Wall Street's willingness to fund money-losing shale operators is, in turn, a reflection of ultra-low interest rates. That poses a twofold risk to shale companies. In his paper for Columbia's Center on Global Energy Policy, Amir Azar noted that if interest rates rose, it would wipe out a significant portion of the improvement in break-even costs.

But low interest rates haven't just meant lower borrowing costs for debt-laden companies. The lack of return elsewhere also led pension funds, which need to be able to pay retirees, to invest massive amounts of money with hedge funds that invest in high yield debt, like that of energy firms, and with private equity firms—which, in turn, shoveled money into shale companies, because in a world devoid of growth, shale at least was growing. Which explains why Lambert, the portfolio manager at Nassau Re, says "Pension funds were the enablers of the U.S. energy revolution."

Speaking in December 2014, just as the bust was beginning, Blackstone CEO Stephen Schwarzman said, "I think this is going to be a wonderful, wonderful opportunity for us. It's going to be one of the best opportunities we've had in many, many years."

92   According to research done by SailingStone, in 2015, nearly $70 billion of capital was raised by private equity funds dedicated to natural resources—a new record. The next year, such private equity firms raised over $100 billion, almost five times the amount that such firms usually raise.

In some cases, private equity firms bought the debt of troubled companies, or provided equity capital for restructurings. They also have acted almost like venture capital firms, providing seed money for entrepreneurs like Campbell and Sellers to assemble land. According to SailingStone, 35 percent of all horizontal drilling today is being done by privately held private equity backed companies.

The private equity titans have made fortunes—but not necessarily because the companies they have funded have produced profits. Some of the returns that the private equity firms have generated have come from selling one company to another, like in the case of Double Eagle, or in taking a company they've funded public. For a long time, the value the public market was willing to accord a fracker was based not on a multiple of profits, which is a standard way of valuing a company, but rather as a multiple of the acreage a company owns. It was a bit like the old dotcom days, when internet companies were valued on the number of eyeballs. The attitude is invest-and-flip, not buy-and-hold.

"I view it as a greater fool business model," one private equity executive tells me. "But it's one that has worked for a long time."

In the summer of 2017, finger-pointing began regarding who was to blame for the red ink in the shale business. True, investors who expect profits are disappointed in shale companies.

Or as Doug Terreson, a one-time petroleum engineer turned top-ranked energy analyst, now with Evercore ISI, asks, "If shale is such a great business, why isn't it creating value for shareholders?" But it turns out shale companies are disappointed in investors too. At a June 2017 investor conference, Al Walker, who is the CEO of Anadarko Resources, the giant Texas based oil and natural gas company, told the assembled crowd, "The biggest problem our industry faces today is you guys."

He went on to explain that the pressure from investors to grow production, regardless of profitability, made it very hard for management teams to do anything but.

There are signs that that is changing. Terreson, who has done extensive research showing that executive pay in this industry is based much more on production growth than whether investors make money, is pushing companies to fix that. Others are, too. "I've been following the energy sector for 15 years, and I've seen the value destruction first hand," says Todd Heltman, a senior research analyst at money manager Neuberger Berman. "Investors are finally forcing change by tying management compensation to shareholder returns. Things are changing, fast. But the industry needs to produce free cash flow to keep attracting our capital." Indeed, an increasingly vocal cadre of investors is arguing that the growth-at-all-costs model doesn't work. It might make executives and their private equity backers rich in the short term, but it doesn't benefit anyone else, especially given that it's not at all clear how limitless the remaining supply of oil actually is. "Our view is that there's only five years of drilling inventory left in the core," one prominent investor tells me. "If I'm OPEC, I would be laughing at shale. In five years,

94 who cares? It's a crazy system, where we're taking what is a huge gift and what should be real for many years, not five years, and wasting it. The industry didn't have stuff to drill for a long time. Now we have it and we're wasting it. We're flooding the market with a low price resource instead of saving it for a rainy day." There are some longtime skeptics who argue that the industry is moving in the right direction. "By focusing on their best acreage and by being very efficient, the better operators have managed to generate moderate levels of positive free cash flow," says Brian Horey, who runs Aurelian Management. "While their rates of return are still below levels that will sustain the industry in the long run, they are trending in the right direction."

But so far, it is just a handful of the better operators. Even as oil prices rose, only five fracking companies managed to generate more cash than they spent in the first quarter of 2018, reported the Wall Street Journal. And so, if the focus on profitability continues, and companies can only drill what they can fund with their own cash flow, will the production be as grand as the forecasters are saying?

To try to get some clarity as to whether the forecasters or the skeptics might be right, I visited Bill Thomas, the CEO of EOG—the so-called "Harvard of Shale"—in his spartan offices high above Houston. It was just after the havoc unleashed by Hurricane Harvey, and although the sun was shining, the city was unusually subdued.

One of the first slides Thomas showed me is one that ranks all the E&P companies by profitability. "EOG One of Few E&Ps Spending Within Cash Flow," it reads. EOG isn't exactly blowing the doors off, but it is making money, which is better than most of its peers. "We are in the business of long-term economic success," Thomas says.

As for the overall industry, he says, "The actual economic returns for investors are lower than some people think." EOG won't drill unless a well can generate a 30 percent internal rate of return at an oil price of $40 a barrel, because after overhead costs, like land and infrastructure, that return gets cut in half—and that's at EOG, which prides itself on keeping overheard costs low. At an oil price of $40 a barrel, Thomas says there is a limited amount of land where it makes economic sense to drill. (The numbers obviously change as the price of a barrel of oil increases, but it is EOG's goal to be profitable throughout price cycles.)

Thomas says that even in the much celebrated Permian, the rock is much more variable than optimists seem to believe, and the "core"—the really good rock—is smaller. This is no small deal. "Better rock responds exponentially better," he says. "It's not linear."

"The Permian has terrified the world oil market but there are overblown expectations of the Permian," he says.

EOG prides itself on its technological expertise, and one of Thomas's pet peeves is the industry-wide idea (which was started by Aubrey McClendon, of course) that shale drilling is a manufacturing business. "They say, 'It's manufacturing mode and you can stamp these things out," he says. "If you stop thinking and starting stamping them out, your results go down. You've got to keep learning." He adds, "There are a lot of startups doing shale, but there's a long learning curve, and it's not that easy. Do not get into manufacturing mode. You can easily mess things up, and if you do, you cannot go back in and fix it."

Perhaps not surprisingly, Mark Papa, whose retirement didn't last long—he has founded a new company called Centennial— echoes some of these sentiments. He argues that the ability of American shale oil to affect the global price of oil, as the most bullish forecasters predict, is not an ongoing phenomenon, but rather a "one-time event" brought about by the confluence of the Bakken, the Eagle Ford, and the Permian hitting the market at the same time. "What's overlooked," he says, "is that shale wells do deplete." Now, he says, the Bakken "is exhausted as a growth vehicle. The Eagle Ford has some power left, but it's not nearly as powerful as it was." Already, he notes, the rig count in the Eagle Ford and the Bakken is falling. By 2020, he says that even in the Permian, the best acreage will be mostly drilled, and after that, he predicts a sizable dropoff. "There is not an endless tsunami of oil," he likes to say.

In a recent report, Goldman Sachs analysts estimated that because of the decline rates of fracked wells, by 2023, it will cost $58 billion in capital investment just to hold production flat.

Indeed, there's an argument that while frackers are recovering more and more oil faster and faster, that might just mean that the wells deplete more quickly. The truth is, no one knows: Fracking hasn't been around long enough for there to be a history, and the Earth's geology is far too complex to give itself up to easy modeling. "I would still describe it as a big science project," an entrepreneur tells me.

Game of Thrones

As 2017 drew to a close, the EIA announced that U.S. net oil imports, including both crude and refined products, were on track to sink to the lowest monthly level since before the Arab oil embargo of 1973. At less than 2 million barrels a day, imports were running well below the apex of more than 14 million barrels a day in the fall of 2005. Exports were at the highest level in American history, roughly twice the previous crude export peak in 1958, and exceeding those of five of OPEC's members.

OPEC wasn't as impervious to lower prices as Ali Al-Naimi, the Bedouin shepherd-turned-oil-minister, seemed to have suggested in the fall of 2014. True, Saudi Arabia and other OPEC members spend a lot less money to get a barrel of oil out of the ground than do U.S. frackers. But it turns out that Al-Naimi's measurement was flawed. That's because the wealthy (and not-so-wealthy) oil-rich states have long relied on their countries' natural resources to support patronage systems, in which revenue from selling natural resources underwrites generous

98   social programs, subsidies, and infrastructure spending. That, in turn, has helped subdue potential political upheavals. "The privileges that Saudi citizens have enjoyed, courtesy of the kingdom's petroleum wealth, have served as a key deterrent against any form of opposition," wrote Jacob Shapiro, the director of analysis at Geopolitical Futures, in late 2015.

This is an expensive way to run a country, and so the patronage system gave rise to the notion of the fiscal break-even price, which essentially is the average oil price that an oil state needs to balance its budget each year. And it really does all come down to oil: McKinsey noted in a December 2015 analysis that Saudi Arabia gets about 90 percent of its government revenue from oil.

As oil prices soared from about $30 a barrel in 2003 to around $110 a barrel from 2011 to 2012, says McKinsey, Saudi Arabia's GDP doubled, making it the world's nineteenth-largest economy by 2014, ahead of Switzerland and Sweden. The easy money, of course, encouraged high spending. According to the Baker Institute for Public Policy, public spending in Saudi Arabia tripled over the decade ending in 2014, resulting in a fiscal break-even price of between $87 and $109 per barrel of oil, up from about $20 per barrel in 2002. That's in part due to the Arab Spring, which led Saudi Arabia to pledge a $130 billion spending package to placate an increasingly restless populace. (The numbers, of course, are estimates, with as many variables as the price of a barrel of U.S. shale. When it comes to obscuring the actual state of affairs, the murkiness of Saudi Arabian finances rivals even the most outlandish U.S. corporate spin.)

As oil prices fell, the spending strategy began to back-fire. According to a Goldman Sachs analysis in June 2017, while OPEC countries have historically been the low-cost producers from both a cost and a budget perspective, the increased spending levels changed everything. Goldman estimates that from 2011 to 2012, OPEC on average needed $10 to $40 *less* a barrel to balance its budget than did Big Oil and U.S. frackers. But by 2016, OPEC needed $10 to $20 a barrel *more* than did the other oil companies.

For a while, policymakers around the world thought the fiscal break-even would set a floor for oil prices. That fallacy was exploded as oil fell to $26 a barrel. And as oil prices plummeted, countries could no longer balance their budgets. The OPEC 2016 yearbook reported that Saudi petroleum exports, which were worth $321.9 billion in 2013, fell to $284.4 billion in 2014—and then to $158.0 billion in 2015, less than 50 percent of the level they'd been at two short years earlier. In its December 2015 analysis, McKinsey said that Saudi Arabia's annual budget deficit was around $100 billion. "Based on current trends, Saudi Arabia could face a rapid economic deterioration over the next fifteen years," wrote McKinsey. "The repercussions of the current period of low oil prices for policymaking, and the unfolding relationship among the state, society, and economy in the Middle East suggest the region is at the most important crossroads it has faced since the initial oil boom during the 1960s and 1970s," according to a Columbia University report.

At the same time, a Game of Thrones was starting in Saudi Arabia. In late 2014, King Abdullah bin Abdulaziz Al Saud, who had led Saudi Arabia for a decade, passed away, leading to the elevation of his brother, King Salman bin Abdulaziz, to the throne. With that came the accession of the new king's son, Mohammed bin Salman. The new deputy crown prince, who has been nicknamed MBS, was then just thirty years old. His father gave him unprecedented control over a huge portion of the economy, including oil. "He's very charismatic, just like Bill Clinton," Dr. Bernard Haykel, a Middle East expert and professor at Princeton, told a panel at Columbia University in early 2017. "If you're with him, you feel like you're the only person in the world. That's why all the coverage is so complimentary." Bloomberg reported that Western diplomats in Riyadh had started to call MBS "Mr. Everything." "In a country long ruled by aging kings, MBS was young, tall, and transparently ambitious," wrote the *New Yorker* in a profile in the spring of 2018.

The changes came as thick and fast as those in Westeros. In May 2016, King Salman replaced oil minister Ali Al-Naimi with Khalid al-Falih, who had been the CEO of Aramco and was said to favor production cuts over Al-Naimi's market share strategy. That fall, the king also replaced the finance minister of twenty years. In 2016 piece entitled "The Fallout from Saudi Arabia's Economic Downslide," Geopolitical Futures wrote, "Appointing new ministers to the two most important Cabinet portfolios shows that the Saudis realize they are facing a historic economic crisis that will be resolved only through transforming the system."

Transforming the system is exactly what Saudi Arabia is trying to do. MBS is viewed as the architect of a breathtakingly audacious plan called Saudi Vision 2030. He has denied that the plan is because of low oil prices. "This vision was going to

be launched, whether the price of oil was high or low," he told
one interviewer. But regardless of the truth of that statement,
there's no question that the kingdom's budget predicament has
dramatically increased the stakes. Among other things, Vision
2030 aims to increase the number of Saudis in private employ-
ment, reduce dependency on oil—MBS said his country had a
"case of addiction to oil"—and most shockingly, sell a stake in
Aramco to the public. MBS told an interviewer that as a result
of this plan, "If oil stopped in 2020, we can live. We need it. We
need it, but I think in 2020, we can live without oil."

It is a high-wire act. The money from the sale, which could
be the largest IPO in history, is supposed to be used to fuel Saudi
Arabia's Public Investment Fund, allowing it to embark on an
ambitious investment plan. (In June 2016, the PIF put $3.5 bil-
lion into Uber, and it has agreed to commit $20 billion to an
infrastructure fund run by Blackstone.) MBS has said that he
expected Aramco to be valued at more than $2 trillion. But the
Wall Street Journal quoted an anonymous Aramco official calling
the number "unrealistic and mind blowing," and said that a
report for investors prepared by consultant Wood Mackenzie
put Aramco's value at around $400 billion.

Few have any real idea what Aramco is worth, because the
company doesn't publicly report its financials. What no one
disputes is that the value is completely leveraged to the price of
oil. And so, in a great irony, a good part of Saudi Arabia's ability
to finance a move away from oil will be due to investors' con-
tinuing belief that oil prices will remain strong.

At any rate, as prices ground lower, the speculation grew
more intense about how long Saudi Arabia could remain on

the sidelines. "As Saudi Arabia refuses to relent . . . it is slowly crushing not only its competitors such as the high-cost OPEC producing nations and marginal U.S. shale companies, but itself as well," wrote "Tyler Durden," a pseudonym for a group of anonymous and acerbic commentators who run the blunt and bearish blog Zero Hedge, in early 2016. "The biggest question is how much longer Saudi Arabia can continue this self-punishment."

There was punishment for the rest of the world, too. By the fall of 2015, a year after OPEC's surprise Thanksgiving decision to maintain production levels, Venezuela was warning of a "catastrophe" if the cartel didn't cut and prop up prices. "OPEC has never been more divided," a longtime oil analyst named Fadel Gheit told CNNMoney. "When prices move by $40 a barrel, stuff in the world starts breaking," one oil man says. "Oil remains the world's most efficient mechanism for translating economic into geopolitical risk," wrote three authors in a lengthy piece called "Fueling a New Order? The New Geopolitical and Security Consequences of Energy," which was published in 2014 by the Project on International Order and Strategy at Brookings. "In the modern era, no other commodity has played such a pivotal role in driving political and economic turmoil, and there is every reason to expect this to continue."

"Picture this," another source says. "It's February of 2016. You are Saudi Arabia. Oil is $26 a barrel. You want consolidation in North America, but you don't want to be blamed for an emerging market contagion that spreads around the world. Would that have happened? Your guess is as good as mine. These things happen in a flash."

And so, two years after the Thanksgiving Day surprise, OPEC and Saudi Arabia threw in the towel and announced the first production cut in eight years.

It wasn't just OPEC. There was a new player involved. According to Bloomberg, in the spring of 2016, MBS had refused to freeze oil production because Iran, which wanted to recover from the sanctions that had been imposed on it, wouldn't participate. "Observers saw it as extremely rare interference by a member of the royal family, which has traditionally given the technocrats at the Petroleum Ministry ample room for maneuver on oil policy," wrote Bloomberg. But Vladimir Putin stepped in.

According to Reuters, the Russian leader played a "crucial role" mediating what had seemed to be an intractable dispute between Saudi Arabia and Iran. This is despite the fact that the Saudis and the Russians are on opposite sides of the wars in Syria and Yemen. Putin, of course, was also eager to see oil prices rise so that he could spur Russia's economy before he faced reelection in March 2018. As for Saudi Arabia, it didn't want to see Russia even more beholden to Iran. It was OPEC's first deal with Russia in fifteen years. The Saudis and Russians agreed to cut production heavily, while Iran was allowed to slightly boost output.

The price of crude oil immediately rose more than 8 percent. At subsequent meetings, OPEC imposed more cuts, and the price steadily rose to around $50 a barrel by the spring of 2017, and then into the $70s by the spring of 2018. Had prices remained in the $30 range, even the most ardent shale enthusiasts don't think there would have been a rebound in U.S. drilling.

104   "If you are a shale oil producer, who brought you back?" asked
Suhail Al Mazrouei, the United Arab Emirates oil minister, at
a conference, reported Bloomberg. "It was OPEC," he answered.
"Without OPEC, there'd be chaos in the market."

It is unclear how long the alliance between OPEC and Russia
will last, or what its breakup would mean for either one—or for
America's shale industry.

But at least for now, shale boosters are claiming vic-
tory. "Above all, it [the production cuts] means that America
has truly reemerged as the world's energy superpower," wrote
Arthur Herman, a senior fellow at the Hudson Institute. "The
Americans have won and the Saudis have lost this crucial round
of the oil war."

A New Era?

"The only certainty is that nothing is certain." So said the Roman scholar Pliny the Elder, and while there is a mountain of scholarship about the geopolitical changes that will be wrought by fracking, his words still speak truth. It isn't just the mysteries remaining in shale itself, and the unstable financial footing of the industry. Even if shale's most ardent believers turn out to be right, and production continues to soar, there are too many variables at work to reduce it all to a simple equation.

For instance, it may be tempting to think of a United States that doesn't need oil from Saudi Arabia or anywhere else, and that supplies gas to much of the world, as a stronger, safer nation, but that may be too facile. Weakening the rest of the world doesn't necessary serve American interests.

U.S. shale has dealt a crushing blow to imports from OPEC's African members, who produce a light crude that is similar to shale oil, and whose economies are heavily dependent on oil. Exports to the U.S. from three of OPEC's African members—Nigeria, Algeria, and Angola—have fallen to their lowest levels in decades. In 2013, Nigerian Oil Minister Diezani Alison-Madueke told CNBC that U.S. shale oil was a "grave concern." And no wonder: Industry publication Hart Energy reported that as a result of the surge of shale oil, several refiners had virtually eliminated their imports of Nigerian oil. And the oil industry accounts for 70 percent of Nigeria's tax revenue. By the summer of 2017, as Nigeria plunged into its biggest economic crisis in years, Alison-Madueke was begging for limitations on U.S. shale oil.

Limiting America's dependence on unstable regions of the world may seem like an unalloyed positive, but the larger effects are quite murky. "It's very, very difficult question as to whether it makes for a safer world or a less safe world," says Norland. "It assuages some economic anxieties, but there are dangers." He worries that economic hardship can lead to increased terrorism and even civil war—a horror anywhere, but one that is especially horrible in countries like Angola, where a civil war that began in 1975 just ended.

Experts do believe that a supply of U.S. oil and a surplus of cheap U.S. natural gas could weaken the country that most aggressively wields its energy supplies as a weapon: Russia. Not only is the U.S. building export facilities, but Europe has invested heavily in LNG import terminals. American LNG is still more expensive than Russian gas due to the cost of liquefaction, and probably always will be. But now there is an option, whereas before, there was none.

As a result, "even if countries don't buy U.S. gas, they can get better terms and more reliable supply" from Rosneft, which is essentially Russia's state energy company, argues Jamie Webster, BCG energy analyst and fellow at the Center for Global Energy Policy at Columbia University. The Baker Institute reports that Rosneft has already had to accept lower prices for its gas; as long-term contracts roll off, there's an argument that Russia will have to make significant concessions on price. "Shale really hurts Russia," says Stephen Arbogast, a professor of finance and the Director of the Energy Center at the Kenan-Flager Business School in North Carolina. "Combine sanctions with shale, and you can put tremendous pressure on the Russian state."

But if the Kremlin's global political ambitions may ultimately be stunted by Russia's faltering finances, there's little proof yet.

In the fall of 2017, the *New York Times* wrote that Russia is "increasingly wielding oil as a geopolitical tool, spreading its influence around the world and challenging the interests of the United States." Among other things, Russia has cut oil deals in the Middle East with Iraq and Libya, and, in late 2016, the Kremlin allowed Qatar's sovereign wealth fund to buy a stake in Rosneft.

Russia has also stepped into China's shoes as the chief financial backer of Venezuela. Over the course of the past three years, the *Times* calculated that Russia and Rosneft have provided Caracas with $10 billion in financial assistance, and have helped Venezuela avoid default on its debts at least twice. In October, Venezuelan President Nicolás Maduro traveled to Moscow seeking fresh financial backing, and thanked Putin "for your support, both political and diplomatic." In return, Rosneft

108   has been awarded licenses to offshore gas fields, as well as a
stake in Citgo, which is a major refiner in the U.S., as collateral
for one of its loans. Rosneft's seizure of that collateral might
crash into the U.S. sanctions put in place to try to punish Russia
for its aggression in Ukraine.

Fracking undoubtedly has made this a more dangerous
path for Russia. Because the low price of oil is causing the
Venezuelan economy to teeter on the brink of collapse, a
default is looking more and more likely. In the fall of 2017,
rating agency S&P declared Venezuela, which had missed several payments on its debt, to be in default. That could leave
Russia and Rosneft holding bad loans that a new government
might not want to pay. Rosneft has said its loans are being
repaid on time.

The shale revolution also figures into America's often
vexing relationship with China, the world's second-largest
economy. China has overtaken the U.S. as the world's biggest
oil importer, but what's even more astounding is that shipments of U.S. crude oil to China, which were nothing before the
lifting of the export ban, hit almost $10 billion in 2017. China
is also on track to become the biggest importer of U.S. LNG; in
early 2018, Cheniere and China National Petroleum Company
signed the first ever long-term LNG delivery contract between
a Chinese company and a U.S. producer. This could help reduce
our trade deficit, and could be a healthy change from a world
where the U.S. and China compete for scare energy supplies.
It may also be a positive change from 2014, when China and
Russia signed a thirty-year deal for Rosneft to deliver $400
billion of gas to China, striking off alarm bells about a close

alliance between two great powers who may be increasingly unfriendly to the U.S.

One downside is that now, there could be a global tit for tat. With the Trump Administration proposing tariffs on some $50 billion of Chinese imports, China could retaliate, and wreak economic havoc here and in global oil markets. The Trump Administration's proposed tariffs on steel and aluminum imports from Mexico and China might complicate even the concept of North American energy independence should those countries find creative ways of retaliating.

The repercussions of shale's rise in the Middle East might be the most complex of all. The lower prices brought about in part by the fracking revolution are rattling OPEC members, Saudi Arabia in particular, and have brought urgency to the critical task of reshaping economies that have long been dependent on the price of oil. The U.S. is importing far less oil from Saudi Arabia than it used to. By early 2018, imports of crude and petroleum products had fallen to 667,000 barrels a day, down from a peak of 2.3 million barrels a day in May 2003.

But what, if anything, this means for future relations between the two countries is unclear. During the Obama Administration, there certainly was some distancing. OPEC's former acting secretary general, Dr. Adnan Shihab-Eldin, has said that Saudi Arabia suffered "extreme disappointment with a number of actions taken by the Obama Administration," including most prominently America's deal with Iran. "The U.S. Saudi relationship has faltered under the Obama administration," stated a policy brief by Rice University's Baker Institute for Public Policy. "Under the surface is a more deep-seated anxiety

110   as the Saudi ruling elite worries that the oil for security bargain is breaking down," wrote Brookings Institution scholars in
"Fueling a New World Order." "There is a real panic—a concern,
a suspicion, a paranoia—in the Middle East that we will extri-
cate ourselves, that we won't be as beholden," a former Obama
administration official told me. "It is there and arguably useful."

And yet the Trump Administration, for all of the President's virulently anti-Saudi message on the campaign trail, and
for all of his talk about energy dominance, has done nothing to
distance America from the desert kingdom. In fact, quite the
opposite has happened. One of Trump's early Oval Office meetings and formal lunches was with Mohammed bin Salman, who
in an interview with the *Washington Post* had praised Trump
as a "president who will bring America back to the right track."
Instead of visiting Mexico or Canada first, as presidents typi-
cally do, Trump went to Saudi Arabia, where he was presented
with gifts including swords, daggers, and cheetah-fur robes,
according to The Daily Beast. During the visit, the Trump
administration announced an arms deal with the Saudis that
is pegged at $350 billion over ten years. (That was a reversal of
the Obama Administration's 2016 policy of blocking some arms
sales to the regime because of civilian deaths in Yemen.)

The moves by the Trump Administration may reflect
incoherence, or they may show that viewing our complicated
relationship with Saudi Arabia solely through the prism of our
need for oil is wrongheaded. There are a lot of reasons it's in
America's interest to have a stable Middle East, whether it's
fighting terrorism, resisting the spread of nuclear weapons,

protecting Israel—or keeping the global economy functioning.
Even if America doesn't need Middle Eastern oil, its allies in
Europe do, and China certainly does. This isn't just altruism.
In a world where over 40 percent of the S&P 500's revenues
come from outside the U.S., the American economy is depen-
dent on the global economy.

This stark truth is laid out in a fall 2016 analysis by Anthony
Cordesman, who holds the Arleigh A. Burke Chair in Strategy
at the Center for Strategic and International Studies in Washington. He analyzed the most recent U.S. Census Bureau data,
and found that six of the top fifteen U.S. sources of imports
are Asian states heavily dependent on Gulf petroleum exports.
For instance, 20.5 percent of U.S. imports came from China,
6.0 percent from Japan, 3.4 percent from South Korea, 2.1 percent from India, 1.9 percent from Vietnam, and 1.7 percent from
Taiwan. These imports include critical components needed by
technology companies like Apple. In other words, to risk Asia's
economy is to risk our own.

If the U.S. were to leave a power vacuum in the Middle East,
it will be filled by someone else, most likely Russia or China.
All of this is why some experts argue that the U.S. military will
keep guarding the region's oil shipping lanes, as it has done for
decades. "Nobody else can protect it and if it were no longer
available, U.S. oil prices would go up," Michael O'Hanlon, a
senior fellow in foreign policy with the Brookings Institution,
told the Wall Street Journal in 2012. He says the U.S. spends $50
billion a year protecting oil shipments—and will continue to do
so. America "could use its dominant naval position and energy

strength as a 'boot on the throat' of China, as some Chinese fear it will," wrote other Brookings Institution scholars in "Fueling a New Order?" "But there would be huge costs. . . . Just ask Vladimir Putin how well it works to wield energy as a stick, rather than as a commodity."

There are compelling arguments that a world of more abundant energy and lower prices will help foster stability in the Middle East and elsewhere. There's an equally compelling argument that precisely the opposite will happen, as states that are dependent on oil revenues face deficits, high youth unemployment, and restive populations.

Certainly, the shifts in Saudi Arabia have continued to shock the world. In June 2017, there was a Red Wedding of sorts when King Salman abruptly elevated his son, Mohammed bin Salman, or MBS, and removed his nephew, Mohammed bin Nayef—who, Bloomberg noted, was "rightly recognized and appreciated . . . for his work fighting terrorism inside and outside the kingdom, and was a key partner of the U.S.—from all his posts.

If there was still a debate about MBS outside of Saudi Arabia, he moved quickly to quell any dissent within his country. In the fall of 2017, Saudi police arrested multiple princes on charges of corruption, including Al-Waleed bil Talal, a well-known investor in many Western companies from Citigroup to Twitter, politicians, and businessmen. Most of those arrested were interrogated not in prison but in Riyadh's Ritz-Carlton hotel, and the detainees reportedly returned hundreds of millions of dollars to the government. "Was it a power grab?" asked *Sixty Minutes* host Nora O'Donnell in an interview with MBS in the spring of 2018. "If I have the power and the king has the power to take action against influential people, then you are already fundamentally strong," responded MBS. "These are naïve accusations."

Tweeted President Trump, "I have great confidence in King Salman and the Crown Prince of Saudi Arabia, they know exactly what they are doing . . ."

At the same time as the corruption crackdown, MBS's reported spending was putting that of the shale kings in perspective, although his taste doesn't seem to run to sports teams. In the fall of 2017, the *Wall Street Journal* reported that he was the buyer of a painting by Leonardo Da Vinci that sold for a record $450.3 million. "Pushing Austerity at Home and Spending Millions Abroad" was the headline of a December 2017 *New York Times* piece, which reported that MBS had also bought a $500 million yacht and a $300 million-plus chateau in France. A spokesman for the royal family has dismissed some of the reports.

In the fall, MBS was the featured speaker at a glittering investment conference in Riyadh that drew the world's most prominent financiers, from Blackstone's Steve Schwarzman to Goldman Sachs's Harvey Schwartz. There, MBS unveiled a plan to create a new city from scratch on Saudi's Red Sea Coast, one that would rely on renewable energy and be staffed largely by robots.

Even as oil prices recovered in 2018, the proposed IPO of Aramco stalled out, with MBS telling *Time* magazine that he was waiting for higher prices "We believe oil prices will get higher in this year and also get higher in 2019," he said. Bloomberg reported that Saudi Arabia wanted prices around $80 a barrel to pay for the government's agenda and to support

114 the IPO. What happens if prices go in the opposite direction is anyone's guess.

At the Columbia panel at which Dr. Adnan Shihab-Eldin spoke, Dr. Steffen Hertog, a Middle East expert who is a lecturer at the London School of Economics, said, "Our prediction error has gone up dramatically . . . both very good and very bad things can happen much more rapidly than was the case from 1962 until January 2015." He summed up his view by saying, "There are a lot of encouraging things, but also a lot of things that could go wrong. We are in a new era."

Make America Great Again

What is obvious, even today, is the enormous impact of shale gas on the domestic economy. "The U.S. has the lowest cost energy prices of any OECD nation," noted the Energy Center's Stephen Arbogast. Prices are hovering at less than half of prices in much of the rest of the world. That means that energy intensive manufacturing in the U.S. "enjoys a significant advantage versus Europe, Japan, or China, all of whom depend upon imported oil and LNG for marginal supplies." John Shaw, the Harry C. Dudley Professor of Structural and Economic Geology at Harvard, said in a 2015 talk, "Nothing has had a more profound impact on the U.S. and global energy economy in the past decade than the emergence of shale gas resources."

This newfound competitiveness is especially obvious in the refining and chemical industries, where new U.S. capacity is being built and exports are at record levels. In 2016, the German industrial trade group BDI even warned that America's cheap natural gas could put European firms at a serious economic

116   disadvantage. "It is [an] enormous advantage for the U.S.," says
EOG's Bill Thomas. "I think LNG is more of a game changer than
oil is, and there is so much natural gas that we can't use it all."
In early 2018, the EIA reported that the US had become a net
exporter of LNG for the first time since at least 1957.

The impact on U.S. manufacturing can be seen in plans such
as those announced in the spring of 2017 by ExxonMobil and
Saudi Basic Industries (a major Saudi Arabian producer of pet-
rochemicals) to invest $10 billion to create a massive plastics
and petrochemical plant in Corpus Christi. Or Big River Steel's
$1 billion steel mill in Osceola, Arkansas. Or Total's $2 billion
ethane cracker in Port Arthur Texas. Or Shell's Pennsylvania
Shell ethylene cracker plant in Potter Township, Pennsylvania,
on the edge of the Marcellus. (Ethane is a feedstock for ethylene;
both are used in manufacturing products like plastic, deter-
gent, and automotive antifreeze.) Or another Shell petrochem-
icals complex that has sprung up along the Ohio River in Beaver
County, Pennsylvania. Since 2010, over three hundred chemical
industry projects worth $181 billion have been announced in
the U.S., according to the American Chemistry Council, a trade
group representing chemical companies.

You might think the clear strength of natural gas could help
create a coherent domestic energy strategy. But it hasn't.

The clearest example of strategic dissonance is the
Trump Administration's monomaniacal focus on bringing
coal back. Or as Trump, who swept the vote of eight of the top
nine coal-producing states, is wont to say: "Putting an end to
the war on coal." Early in his administration, he was flanked

onstage by more than a dozen coal miners as he announced
plans to review one of President Obama's signature efforts,
the Clean Power Plan, which mandated limits on emissions
from power plants. The miners "told me about the efforts to
shut down their mines, their communities, and their very way
of life," Trump said. "I made them this promise: We will put
our miners back to work." As if to refute McClendon's "Coal is
Filthy" campaign, Trump calls it "beautiful clean coal."

It's certainly true that coal is in decline. Last year, U.S.
demand for coal fell almost 10 percent, the biggest decline in
the world in absolute terms, according to data from BP. Since
2008, four of the country's big coal companies have gone out of
business.

But this isn't so much a result of government policy, let
alone a "war on coal," as it is of market forces—market forces
that the Trump Administration is, in other aspects, celebrating
and encouraging by its actions elsewhere. Cheap natural gas
is what's destroying coal, not tree-hugging liberals. It's now
cheaper to build a power plant that uses natural gas than to
build one fueled by coal. As a result, the market share for coal in
power generation has fallen from 50 percent in 2005 to around
30 percent today.

In 2017, multiple coal-fired plants have announced their
closings, from PSEG's last two coal-fired plants in Jersey City,
New Jersey, to We Energies' thirty-five-year-old Pleasant
Prairie plant in Kenosha, Wisconsin. At the same time, there
are nine gas-fired plants either under construction or in
some phase of development in Ohio, most of which are near

118   the shale fields in eastern Ohio. "Natural gas will replace coal as the greatest fuel component of the world's total primary energy required by 2040," predicts the Baker Institute.

The environmental impact is probably positive. In large part because of the switch to natural gas, U.S. carbon dioxide emissions from energy sources hit a twenty-five year low in 2017. But at the same time, there's an argument, backed up by a 2017 NASA study, that a huge rise in emissions of methane, which some scientists argue is a more potent greenhouse gas than carbon dioxide, is due to the natural gas supply chain—which could negate some of the environmental benefits of natural gas. Among the open questions is to what extent the industry can control leaks if it chooses or is forced to do so.

Instead of focusing its efforts on forcing improvement in the natural gas supply chain, the Administration has thus far been trying to bail out the coal industry. In the fall of 2017, the Department of Energy, under Rick Perry, announced a plan that would in effect force regional electricity grids to purchase large amounts of coal. The ostensible reason is to ensure a supply of fuel that can be stored and called upon in the event of a disruption—but in addition to distorting markets and likely causing an increase in energy prices, the plan ran counter to the Department's own study, which reported that increased reliance on natural gas and renewables was not reducing the reliability of the grid. Experts promptly dismantled the argument that coal has some kind of magical power; after all, soaking wet or frozen piles of coal haven't exactly saved us during past interruptions, and there might be a more modern way to shore up the security of the grid than stockpiling coal. Former FERC commissioner

Nora Mead Brownell told a trade magazine that the proposal is "the antithesis of good economics. It's going to destroy the markets [and] drive away investment in new, more efficient technologies."

What the plan did was create a strange set of allies. Oil and gas companies joined solar and wind advocates in working aggressively against it. The American Petroleum Institute warned the Trump Administration that it had better not hurt natural gas in an effort to help coal and nuclear energy. A coalition of eleven energy associations, from the Natural Gas Supply Association to the Solar Energy Industries Association, sent a letter asking to delay the implementation of the rule. "This seems to be putting the thumb on the scale against natural gas," said Dena Wiggins, the president of the Natural Gas Supply Association. Bob Flexon, the CEO of Dynegy, called it a "cannon aimed at natural gas."

It's not just America. In some quarters, the fear is that without U.S. leadership, the clock in the rest of the world will turn back to coal. "From a pure economic standpoint, the debate is over here," says Webster. "But I talk to big private oil and gas CEOs, and they are really worried that other countries, where coal is still cheaper, will default to coal." Once a power plant that uses coal is built, that locks in its use for decades to come.

In early 2018, the Federal Energy Regulatory Commission unanimously rejected Perry's plan, despite the fact that four of the five commissioners who lead the agency were appointed by President Trump. Subsequently, the *Washington Post* reported that the Trump Administration was preparing "more drastic alternatives" that essentially would force the purchase of coal.

120 (The Energy Department has disputed the notion that it was trying to find a backdoor need to bail out the coal industry, and has cited the need for a redundant supply to make the grid secure.)

The Administration is also trying to use the EPA, run by Oklahoman Scott Pruitt, to save coal. Among other things, the EPA has rolled back the implementation of Obama Administration rules to control emissions that threaten human health.

At the same time that the Trump Administration proposed a plan that might reduce the demand for natural gas, they are also talking about unleashing more supply. Trump's initial proposal was to unlock our "$50 trillion in untapped shale, oil, and natural gas reserves," and to further that goal, he signed an executive order to ease regulations on offshore drilling and eventually allow more to occur, particularly in the Arctic Ocean. His administration has also proposed allowing drilling in the Arctic National Wildlife Refuge for the first time in forty years. The effect of all of this would be to make it harder to control methane leaks, and it would also likely crater prices, thereby making the economics of drilling even less attractive than they already are. If this comes about at a time of rising interest rates and the end of the era of cheap capital, we may soon begin talking about how the Trump Administration killed the shale revolution.

Losing the Race

Geology is the one thing the shale revolution, whatever shape it takes in the future, cannot change. Fossil fuels are called non-renewables for a reason: Once they are gone, they are gone. At some date in the future, the world may have extracted everything there is to extract. Renewables like wind, solar, and water power will either have replaced them, and the debate between David Einhorn and the oil drillers about the economic sustainability of fracking will have receded into the past—or they won't, and we will all be in trouble.

In 2011, Aubrey McClendon told *Forbes*, "The reality is that wind and solar can never be more than about 15 percent of our power requirements and will likely never be cost competitive with natural gas." Maybe that's the way it looked from the vantage point of the "Fracking King," as Bloomberg once called McClendon. But if it looked that way once, it no longer does.

In conversations with several large private equity investors, I was stunned to hear that they were no longer investing in

122   oil and gas—not so much because of ethical concerns about the environment, but rather for the simple reason that they didn't think the profits would be there for much longer. The end of oil is coming, and all the market has to do is see the end for the price to plunge. "The biggest risk [to the oil and gas industry] is demand destruction," says Steve Wood, a managing director for the oil and gas team at rating agency Moody's. The price for renewables is falling rapidly, to the point that management consulting firm Arabella Advisors predicts that by the end of this decade, solar power will be cheaper than fossil fuel power—without subsidies.

What no one knows is the timing. "The multi-billion dollar question is the pace of renewables versus fossil fuel," noted another major energy private equity investor, who even convened a large meeting of experts and investors in an ultimately unsuccessful effort to arrive at an answer. "You can see the reason why this matters with thermal coal [coal used for power generation]. The minute the market started to see that it was moving into a permanent decline, the prices never recovered."

Once the price goes into decline, it will be far harder, and far more expensive, for oil and gas companies to raise money— especially America's independent oil companies. "These companies are competing against each other, but they are really competing against Petrobras, Aramco, and Rosneft," says Wood at Moody's. "They are competing against countries. And the countries will be the ultimate survivors in a declining oil demand world."

How fast this change comes about is dependent on a host of variables, the outcomes of which rest on yet more variables, creating a multifactorial puzzle, the solution to

which no one truly knows. The variables range from the rate
of growth in electric passenger vehicles to the pace of battery
development to the development of technology that will allow
renewables to be used to fuel commercial transportation, like
trucks and jets. In October 2016, Fitch Ratings, a leading credit
rating agency, called widespread adoption of battery-powered
vehicles "a serious threat to the oil industry," noting that
battery costs have fallen by 73 percent since 2008 and electric cars are nearing cost competitiveness with gas- and
diesel-powered vehicles. Policy will play a role, too. Before the
Paris Agreement, there were already more than eight hundred
climate change laws on the books around the globe—ranging
from carbon taxes to clean energy investment mandates. Will
there be more mandates? China, the world's largest automotive market, says it is working on a timetable to implement a
ban on vehicles powered by fossil fuels.

Most scholars think the transition will take decades. But
there are those who say it might come much more quickly. Most
prominently, in a 2015 study, Stanford University engineering
professor Mark Jacobson and colleagues have argued that it was
technically feasible for all fifty states to run on clean renewable energy by 2050, with an 80 percent conversion possible by
2030. Of course, there's also the opposite argument. Fatih Birol,
the executive director of the International Energy Agency, a
Paris-based nongovernmental organization, doesn't think we'll
see peak demand for oil anytime soon, mainly because very few
countries have any sort of fuel economy standards for trucks, and
the use of renewables in freight transportation lags passenger
vehicles enormously. And while passenger cars make up about 25

124   percent of oil demand, other modes of transportation, like shipping, aviation, and freight account for almost 30 percent. The IEA says that oil demand from road freight is projected to grow by 5 million barrels per day by 2050, or around 40 percent of the projected increase in global oil demand in that period. OPEC too projects that demand will keep increasing through 2040.

But just because no one knows the answer doesn't make it smart to pretend that the time isn't coming. Which, these days, seems to be America's strategy. President Trump has proposed slashing the budget for a division of the Department of Energy called the Office of Energy Efficiency and Renewable Energy, which is tasked with the development of clean energy like solar and wind power. The proposed spending cuts caused the last seven heads of the office, including three who served under Republican presidents, to write a letter to Congress. "We are unified that cuts of this magnitude . . . will do serious harm to this office's critical work and America's energy future," they wrote. Trump has also imposed tariffs on foreign-made solar panels, which could decrease installation volumes in coming years; Bloomberg called the tariffs "the biggest blow to renewables yet."

In the tax bill that was wending its way through Congress as 2017 drew to a close, Axios's Amy Harder reported that the nonpartisan Taxpayers for Common Sense had tallied up nearly $50 billion of subsides going mostly to the fossil fuel sector that were kept intact in the bill. Subsides for wind and solar, the ones that were extended in exchange for allowing oil exports, were kept too—but those are set to expire in coming years, and Environmental Protection Agency chief Scott Pruitt has called for ending them.

While the Trump Administration plans to effectively subsidize coal, America continues to fall behind in the race to develop renewables. "Why China is Winning the Clean Energy Race" was the headline of a recent Axios piece, which pointed out that Beijing is close to launching a national cap-and-trade program that is slated to be the most sophisticated carbon market in the world, and that China has blown past the U.S. in deploying new clean technologies. Part of the reason, the article argued, is that unlike China, the U.S. has no long-term, coordinated energy and environmental strategy.

According to a Bloomberg report entitled *Global Trends in Renewable Energy*, China is now the leading destination for renewable energy investment, accounting for 45 percent of the global total in 2017. In contrast, investment in renewable energy in the U.S., which was already well below China, declined last year.

China certainly isn't alone. As America celebrates its supposed crushing of OPEC, the Middle East is planning for the future. Saudi Arabia, for instance, is planning not just to build Saudi crown prince MBS's new city powered by renewable energy, but also to spend $50 billion on a massive push into solar. The country just received bids from Abu Dhabi's Masdar and Electricite de France SA to supply the cheapest solar electricity ever recorded. Even Saudi Arabia is trying to move its economy away from fossil fuels in order to generate as much money as they can from exporting oil—until the day it's all over. "In twenty years, oil goes to zero, and then renewables take over," MBS told a gathering of venture capitalists in San Francisco recently, according to the New Yorker. "I have twenty years to reorient my country."

Epilogue

I began this book because I was uncertain about the consequences of fracking. I wanted to ask the questions that were not being asked, because I was skeptical about the "shale revolution," rosy claims of American energy independence, and how it would restore the country's depleting geopolitical power. I found that it's a fool's errand to make bold predictions about what's to come, but the most honest answer I found about the future came from research firm IHS Markit.

The firm has three scenarios. The first, called Rivalry, is the base case. Rivalry, IHS says, means "intense competition among energy sources plus evolutionary social and technology change. Gas loosens oil's grip on transport demand. Renewables become increasingly competitive with gas, coal, and nuclear in power generation."

The second scenario, called Autonomy, is a much faster-than-expected transition away from fossil fuels. "Revolutionary

changes in market, technology, and social forces decentralize
the global energy supply and demand system."

The last scenario is called Vertigo. Vertigo means "economic and geopolitical uncertainty drive volatility and boom-bust cycles with economic concerns slowing the transition to a less carbon-intensive economy."

Under IHS's Autonomy scenario, we face a lot of risks to our future if we aren't developing renewables. But we, along with the rest of the world, probably face the greatest risks under the Vertigo scenario.

The potential for Vertigo helps explain why Charlie Munger, the famous investor and thinker and longtime Warren Buffett sidekick, believes that we should conserve what we have, instead of drilling frenetically. Munger argues that for all the eventual certainty of renewables, there is still no substitute for hydrocarbons in several essential aspects of modern life, namely transportation and agriculture.

Part of why the U.S. can feed its population is because yield per acre has increased dramatically in the modern era—but that's in large part due to pesticides, nitrogenous fertilizers and other agricultural products that are made through use of hydrocarbons. "We need to eat as far out as you can see," he says. "It is a serious problem, and I think hydrocarbons as chemical feed stock are probably irreplaceable. To do well without hydrocarbons requires a new technology that doesn't yet exist. For now, hydrocarbons are like the topsoil of Iowa. You wouldn't want to use it as fast as possible. You would want to use it as slowly as possible. People think there will be a replacement for hydrocarbons, but I don't think that's a safe assumption."

In his view, the U.S. has gotten lucky to find a huge new store of shale oil and gas—America should keep it in reserve unless or until it's clear that it won't need it. "Imported oil is not your enemy, it's your friend," he said at a U.S.–China relations conference. "Every barrel that you use up that comes from somebody else is a barrel of your precious oil which you're going to need to feed your people and maintain your civilization. You want to produce just enough so that you keep up on all of the technology. And you shouldn't mind at all paying prices that look high for foreign oil." He adds, "You will be better off because you delayed gratification, instead of grabbing for it like a child."

Another way to think about this is that America is the only country in the world to have made the switch to unconventional oil and gas—America is the only country to have exhausted its supplies of conventional oil and gas. And other countries have shale, too. Thus far, high profile projects in Poland and China have met with a total lack of success, for a wide range of reasons, from the different quality of the rock to a lack of transportation infrastructure needed to move the supply. But it's probably safe to say if there were an absolute need, they would get it out. "American exceptionalism allowed us to move first," says Webster. "But given time and incentives, other countries will figure it out if they need to."

As history shows, even oil and gas executives don't have a clue what's going to happen next. Charlie Munger might be right. Or

shale oil and gas might do what shale oil and gas have done since
the revolution began, and surprise to the upside. EOG might
discover ways to get oil economically out of other places we
never thought we could get oil economically. Or there could be a
battery breakthrough tomorrow that renders oil obsolete more
quickly than anyone ever dreamed.

In the face of that, there a few things that should begin to
change. For one, we should recognize that America's oil and gas
resources are very different. America's natural gas is ultra-low
cost, and even conservative estimates say that the supply should
last for at least a century. Oil is very different. In oil, neither the
price at which the oil is recoverable nor the ultimate supply are
clear. All we know is that other countries have far more of it, and
they can recover it for far less.

We should also have a degree of humility about the extent
to which both oil and to a lesser extent gas drillers are dependent on the willingness of the capital markets to finance them—
and perhaps a plan for what to do if, in the face of higher interest
rates, that changes. For the first time in perhaps forever, at
least some long-term investors are aligned with conservation-
ists, and they are trying to send a message that isn't drill, baby,
drill—but rather drill thoughtfully and profitably, so that more
people benefit from America's resources for longer, and it isn't
only executives getting a payday.

The great irony may be that Aubrey McClendon got some
important things right. After his death, a journalist named
Martin Rosenberg recounted how he'd asked McClendon what
our national energy policy should be. "Embrace natural gas to

reduce our importation of oil and embrace natural gas to reduce our consumption of coal," McClendon responded. In a recent letter to the firm's clients, JP Morgan chief strategist Michael Cembalest wrote that one thing he considered critical for our future was "the ability to develop natural gas-powered vehicles and trains with lower fuel costs than gasoline- or diesel-powered counterparts, and with greater geopolitical fuel security."

Cembalest also noted another truth, which is that renewables like wind and solar probably will not be entirely sufficient on their own, for the simple reason that they can't be stored. The energy grid of the future will likely consist of mostly renewables, but with the ability to rapidly add backup power from natural gas when wind, solar, and hydropower generation is low. As Cembalest wrote, "An electricity grid with less coal, less nuclear, and more renewable energy would be highly dependent on abundant, low-cost natural gas."

The capacity to export is still a good thing, because the real possibility of American supply can be as much of a stick in geopolitics as the actuality of American supply. It doesn't make sense to shoot for the false notion of energy independence, much less dominance. Like it or not, the U.S. will be part of the world. The challenge is to find ways to use domestic resources to support America's broader geopolitical goals, not as an end in themselves, and not as a threatening stick.

In 1975, when President Gerald Ford signed the Energy Policy and Conservation Act that banned oil exports, the *New York Times* wrote, "Institutionally, Congress seems incapable of examining the 'energy independence' issue in its broadest aspects and of writing an integrated energy policy that is internally consistent and consistent also with economic, foreign policy, and environmental goals."

For all the profound changes in the energy world since then, that, unfortunately, has remained a constant.

The Frackers: The Outrageous Inside Story of the New Billionaire Wildcatters by Gregory Zuckerman. This is the in-depth, definitive story by a longtime *Wall Street Journal* reporter of how wildcatters, from George Mitchell to Harold Hamm of Continental, made fracking into a reality, and made fortunes for themselves in the process.

The Boom: How Fracking Ignited the American Energy Revolution and Changed the World by Russell Gold. Gold's book is also an account of the birth of fracking, but he takes a personal and environmentally minded view. Gold, who also writes for the *Wall Street Journal*, is an exceptionally fair and factual reporter, and the book is clear about the pros and cons—as well as the reasons why fracking has become such a battleground.

The Green and the Black: The Complete Story of the Shale Revolution, the Fight over Fracking and the Future of Energy by Gary Sernovitz. Sernovitz, a one-time aspiring writer turned financier and venture capitalist, has written a fast-paced and engaging book that answers all the questions you could ask about the risks and benefits of fracking.

Windfall: How the New Energy Abundance Upends Global Politics and Strengthens America's Power by Meghan L. O'Sullivan. O'Sullivan, who is a former deputy national security adviser on Iraq and Afghanistan and is now the director of the Geopolitics of Energy Project at Harvard, is thoroughly steeped in all the ways in which America's newfound oil and gas wealth may change the world. Her book, which assumes a continued bounty but nevertheless is skeptical of the notion of "energy independence," is a deep dive into future possibilities.

Fueling a New Order? The New Geopolitical and Security Consequences of Energy by Bruce Jones, David Steven, and Emily O'Brien. This piece is a relatively quick but nonetheless very nuanced look at the ways in which our energy abundance has and will continue to affect geopolitics.

134   "A Saudi Prince's Quest to Remake the Middle East" by Dexter Filkins in
        the *New Yorker.* This profile of Mohammed bin Salman, otherwise known
        as MBS, is a close look at the charismatic and very young man who is
        likely to have an enormous impact on the Middle East—and on us—in
        the coming years.

NOTES

INTRODUCTION

13 **an oil powerhouse:** "United States Remains the World's Top Producer of Petroleum and Natural Gas Hydrocarbons," U.S. Energy Information Administration, May 21, 2018.

16 **"fracking" is viewed by the industry as a pejorative:** Style Guide, Colorado School of Mines.

17 **real catalyst of the shale revolution:** "Reserve Base Lending and the Outlook for Shale Oil and Gas Finance," Amir Azar, Center on Global Energy Policy, Columbia University, May 3, 2017.

CHAPTER ONE

20 **party boys and geniuses:** *The Frackers: The Outrageous Inside Story of the New Billionaire Wildcatters,* Gregory Zuckerman, Penguin, 2013.

23 **wrangling over the remains of his estate:** "Was Aubrey McClendon a Billionaire, or Broke?" Ryan Dezember, *Wall Street Journal,* March 2, 2017.

24 **the biggest blowout:** "The Big Fracking Bubble: The Scam Behind Aubrey McClendon's Gas Boom," Jeff Goodell, *Rolling Stone,* March 1, 2012.

31 **worst period of our careers:** "All Gassed Up," Nissa Darbonne, *Oil & Gas Investor,* September 2006.

32 **$18 billion over the five years:** "Projecting the Economic Impact of the Fayetteville Shale Play for 2008–2012," Center for Business and Economic Research, University of Arkansas, March 2008.

33 **fraternity boy what to do with the beer:** "Independents Encouraged to Spend More on Exploration," *Natural Gas Intelligence,* February 21, 2005.

CHAPTER TWO

40 **one-bedroom apartment in Williston:** According to notes circulated after a meeting of the North Dakota Sheriff & Deputies Association.

41 **Eagle Ford contained over nine hundred million barrels of oil:** Zuckerman.

42 **71 requests for drilling:** *The Boom: How Fracking Ignited the American Energy Revolution and Changed the World,* Russell Gold, Simon & Schuster, 2014.

CHAPTER THREE

46 **Lavish and Leveraged Life:** "Special Report: The Lavish and Leveraged Life of Aubrey McClendon," John Shiffman, Anna Driver, Brian Grow, Reuters, June 7, 2012.

48 **debt on its balance sheet:** "The Incredible Rise and Final Hours of Fracking King Aubrey McClendon," Bryan Gruley, Joe Carroll, and Asjylyn Loder, *Bloomberg Businessweek,* March 10, 2016.

136   49   **existence of the billion-dollar-plus loans:** "Exclusive: Chesapeake CEO Arranged New $450 Million Loan from Financier," Jennifer Ablan, Reuters, May 8, 2012.

51   **print out a map of acreage:** *Chesapeake Energy Corp. v. American Energy Partners LP et al*, District Court of Oklahoma, February 17, 2015.

CHAPTER FOUR

54   **Mother Fracker:** According to a presentation slides and a speech given at the Sohn Investman Conference by David Einhorn on May 4, 2015.

55   **average well in the Bakken declines 69 percent:** "What Could Lower Prices Mean for U.S. Oil Production?" Nida Çakır Melek, Federal Reserve Bank of Kansas City, 2015.

55   **net debt exceeded $175 billion:** Azar.

57   **myopic obsession with production growth:** According to a letter SailingStone Capital sent investors.

CHAPTER FIVE

59   **first job when he was four years old:** *Out of the Desert: My Journey From Nomadic Bedouin to the Heart of Global Oil,* Ali Al-Naimi, Penguin, 2016.

61   **Saudi production costs are no more than $5 per barrel:** "MEES Interview With Ali Naimi: 'OPEC Will Never Plan To Cut,'" *Middle East Economic Survey*, December 22, 2014.

62   **catastrophic:** "Moody's: 2015 E&P Sector Defaults, Creditor Losses on Track to Significantly Exceed," Moody's, September 12, 2016.

64   **ambitions are fading fast:** "Sweetwater Prepared for Book, Now Awaits Bust," Emily Schmall, The Associated Press, January 9, 2015.

CHAPTER SIX

66   **a hundred financing statements:** "Aubrey McClendon Bet Big to Finance Second Act," Ryan Dezember, Bradley Olson, and Erin Ailworth, *The Wall Street Journal*, March 7, 2016.

68   **bash each other's brains out:** "Encana Settles Michigan Antitrust Case, Chesapeake Fights On," Brian Grow, Reuters, May 5, 2014.

70   **private dinner with potential business partners:** "Special Report: The Final Days and Deals of Aubrey McClendon," John Shiffman, Brian Grow, and Michael Flaherty, Reuters, March 11, 2016.

70   **wrong and unprecedented:** "Aubrey McClendon Responds to (But Doesn't Deny) Federal Conspiracy Charges," Christopher Helman, *Forbes*, March 1, 2016.

CHAPTER SEVEN

75   **Saudi America:** "Oil: The Next Revolution," Leonardo Maugeri,

Belfer Center for Science and International Affairs, Harvard Kennedy School, June 2012.

75 **significant gross exporter:** "The Case for Allowing U.S. Crude Oil Exports," Blake Clayton, Council on Foreign Relations, July 8, 2013.

78 **refiners even formed their own lobbying group:** "Independent Refiners Join to Fight Crude-Export Plans," Jennifer A. Dlouhy, *Houston Chronicle*, March 17, 2014.

80 **extensions on tax credits for wind and solar power:** "Congress Passes $1.15 Trillion Spending Bill," Kristina Peterson, *Wall Street Journal*, December 18, 2015.

CHAPTER EIGHT

84 **God felt such remorse:** "That's Oil, Folks!" Skip Hollandsworth, *Texas Monthly*, September 2010.

85 **living in tents, cars and trailers:** "Midland, TX," John Leffler, *Handbook of Texas Online*.

88 **rigs drilling more wells:** "Unraveling the Oil Conundrum: Productivity Improvements and Cost Declines in the U.S. Shale Oil Industry," Ryan Decker, Aaron Flaaen, and Maria Tito, Federal Reserve, March 22, 2016.

90 **industry has burned up cash:** "America's Shale Firms Don't Give a Frack About Financial Returns," *The Economist*, March 25, 2017.

CHAPTER NINE

98 **90 percent of its government revenue:** "Saudi Arabia Beyond Oil: The Investment and Productivity Transformation," McKinsey Global Institute, December 2015.

100 **long ruled by aging kings:** "A Saudi Prince's Quest to Remake the Middle East," Dexter Filkins, *The New Yorker*, April 9, 2018.

102 **anonymous and acerbic commentators:** "As The Saudi Economy Implodes, a Fascinating Solution Emerges: The Aramco IPO," Tyler Durden, Zero Hedge, January 7, 2016.

102 **translating economic into geopolitical risk:** "Fueling a New Order? The New Geopolitical and Security Consequences of Energy," Bruce Jones, David Steven, and Emily O'Brien, Brookings Institution, April 15, 2014.

CHAPTER TEN

105 **crushing blow to imports:** "Nigeria Bearing Brunt of U.S. Shale-Oil Boom," Sarah Kent, *The Wall Street Journal*, March 6, 2013.

111 **swords, daggers, and cheetah fur robes:** "The Insane Gifts Saudi Arabia Gave President Trump," Ken Klippenstein, The Daily Beast, September 4, 2017.

110 **Asian states heavily dependent on Gulf petroleum:** "President Trump's Trip to Saudi Arabia," Anthony H. Cordesman, Center for Strategic and International Studies, May 11, 2017.

138   112   **"Was it a power grab?":** "Saudi Arabia's Heir to the Throne Talks to *60 Minutes*," Norah O'Donnell, *60 Minutes,* March 19, 2018.

CHAPTER ELEVEN

118   **EPA has rolled back:** "67 Environmental Rules on the Way Out Under Trump," Nadja Popovich, Livia Albeck-Ripka, and Kendra Pierre-Louis, *The New York Times,* October 5, 2017.

119   **rise in emissions of methane:** "Reduced Biomass Burning Emissions Reconcile Conflicting Estimates of the Post-2006 Atmospheric Methane Budget," John Worden, *Nature Communications,* December 20, 2017.

CHAPTER TWELVE

121   **15 percent of our power requirements:** "In His Own Words:

Chesapeake's Aubrey McClendon Answers Our 25 Questions," Christopher Helman, *Forbes,* October 5, 2011.

123   **powered by wind, solar, and water:** "100% Clean and Renewable Wind, Water, and Sunlight All-Sector Energy Roadmaps for 139 Countries of the World," Mark Jacobson, *Joule,* September 6, 2017.

EPILOGUE

126   **three scenarios:** "IHS Long-Term Planning and Energy Scenarios," IHS Markit.

129   **Embrace natural gas:** "Remembering McClendon," Martin Rosenberg, The Energy Times, March 8, 2016.

130   **letter to the firm's clients:** Michael Cembalest, J.P. Morgan, Annual Energy Paper, June 2017.

Columbia Global Reports is a publishing imprint from Columbia University that commissions authors to do original on-site reporting around the globe on a wide range of issues. The resulting novella-length books offer new ways to look at and understand the world that can be read in a few hours. Most readers are curious and busy. Our books are for them.

Pipe Dreams: The Plundering of Iraq's Oil Wealth
Erin Banco

Never Remember: Searching for Stalin's Gulags in Putin's Russia
Masha Gessen and Misha Friedman

High-Speed Empire: Chinese Expansion and the Future of Southeast Asia
Will Doig

The Nationalist Revival: Trade, Immigration, and the Revolt Against Globalization
John B. Judis

The Curse of Bigness: Antitrust in the New Gilded Age
Tim Wu

We Want to Negotiate: The Secret World of Kidnapping, Hostages and Ransom
Joel Simon